I0491660

WAKE UP! IT'S JUST A DREAM. THE NIGHTMARE OF LIFE.

Joseba Gorka Isasi Sainz

E-mail j_isasi@hotmail.com

Dedicated to my mother

Synopsis

In this work I have tried to bring together various thoughts that have been accumulating over the years in one place. As I have been moving forward with my notes and memories, I have been putting internal order at the same time, and the result is the book you now hold in your hands.

I deal here with very diverse, controversial and, I think, topical subjects; simulations, quantum mechanics, dreams, astronomy, meditation... I believe that some subjects are not given the attention they should be given and we continue to obsess over banal subjects to which we are dragged with our daily chores.

This work is not intended to be a complete compendium of the various topics covered, but rather a preamble with an imperative need, from my point of view, to deepen in a general and also private way, since, for some topics, there is no other way out but to take the path indicated independently.

The objective therefore would be to try to open the eyes of the public in general, and you in particular, since you have decided to read this text. To open the eyes to submit all the established concepts to a critical judgment and start thinking for ourselves without being bound by what is dictated in a general way, which in many cases, is nothing but castles on clouds that should fall under their own weight.

Without further ado, open your eyes and criticize me openly to everything I expose, since this will undoubtedly enrich me.

The manuscript of this book was started on November 20, 2019 and concluded on or about July 31, 2020.

Index

Illustrations

1. Simulation

Who hasn't seen the movie Matrix? In this movie, the protagonist Thomas A. Anderson is a computer programmer by day and a hacker, named Neo by night. He has been sensing all his life that there is something else, that something is wrong, that something is missing...

He receives a message: "Matrix possesses you", and with the development of the film, with the help of other hackers, Morpheus and Trinity, he arrives at the crossroads of discovering what Matrix is by taking a red pill and renouncing his current life, or, on the contrary, taking a blue pill and continuing as he was. He chooses the red pill to discover that the world he knows is a virtual simulation in which he is connected by a cable plugged into his brain. All around him, billions of equally connected people are being cultivated to power the machines.

Let's look at two points here:

- Neo senses that there is something else, that something is missing; and

- Neo takes the red pill, i.e., he opts for the unknown path.

Aren't these a couple of characteristics that define us? Aren't we always looking for something else to fulfill us, to satisfy us?

And isn't it a human trait to seek adventure, to discover the virgin path, or at least the path we have not gone before?

Whatever we do, we are always dissatisfied, and many times we are capable of doing the unimaginable to try to satisfy ourselves.

So, just as Neo lived in a simulation, who can tell us that we ourselves do not live in a simulation and that all our anxieties are the fruit of that simulation?

Let's imagine for a moment that this is the case, i.e., that we are inside a simulation. What does this mean?

First of all, our perspective changes: we are no longer the masters of our lives, and being a simulation, we take for granted that there is something or someone in control of the simulation.

Who or what controls this possible simulation?

There are countless answers here. We could be ourselves in a more developed society, where we have become so dependent on the very comforts that we have been able to invent, that we are now unable to move and are basically lying on sofas constantly fed and drained through tubes, where we have plugged ourselves into the simulation to make our pitiful existence more entertaining.

In this sense, the purpose of the simulation would be to entertain, to amuse. As if we were watching a movie, or a series. And I am not saying SERIES with capital letters, but a series, one of the many possibilities available.

This series, our simulation, our life, might seem uninteresting and doomed to failure, but seeing the commercial success of Big Brother where the only thing that is done is to see the lives of other people, it could even be a great success.

Within this same branch of entertainment, we could also be inside a video game, just like in another movie with the title Real Player One, although in this situation, I do believe that the simulation as it is, would be a video game of little affluence...

Simulation could also be an experiment where again something or someone poses a series of scenarios and lets them evolve to see what happens and draw conclusions.

Again, it could be something controlled by ourselves in a more evolved future, where we want to see what would happen in the world if a character like Trump rules in the United States.

Being an experiment, as in all experiments, our simulation, our world would be stuck in a test tube in a laboratory full of test tubes with different starting conditions: billions of parallel universes.

Another clear possibility would be that of the Matrix itself; we would be like batteries for the machines we ourselves created one day and which eventually overtook us. Just as we enslave cows, chickens and other animals to live and die in tiny cubicles while we do not stop milking milk and requiring eggs to end up being cut up and taken to the frying pan, so we could be in a similar situation, with the only difference that something or someone kinder than us, has created the simulation for us to live with some comfort in our fantasy.

Instead of machines in the previous scenario, the masters could also be the human beings themselves in a future where the differences between the rich and the poor have grown so large, and where technology has advanced so much, that the rich no longer need cheap labor, nor the problems that can give that sector of poor society, so they decide to put the poor in cabins putting them the television called simulation while they take advantage of what they deem appropriate: energy, organs, ...

Basically, that is something that is already being done, albeit on a different scale, with soccer, discussions at the Big

Brother level... while serious problems such as climate change take a back seat without importance.

The controller could also be an extraterrestrial society that has conquered us and now has us at its service.

Maybe even our planet Earth is not really our planet, and maybe we don't even have the physical appearance we think we do, with arms and legs, and we are some kind of worms living the simulation.

In any case, and at a less speculative level, assuming that what we experience is real, we can affirm without a doubt that what we think and believe to be real is only a representation of what is probably outside our brain.

We must not forget that what our senses perceive - sight, hearing, taste, touch and smell - is transmitted to our brain, and our brain interprets. The interpretation between your brain and mine may be different under the same stimulus. Even if the interpretation is very similar, there will always be small connotations that make the points of view diverge. These connotations can be caused by feelings and emotions of the instant, by memories, or even simply because we are not able to have you and me occupying the same place at the same instant, which makes our perceptual angles differ.

With this first chapter, I want to say that, although we are not living in a simulation as it has been interpreted at the beginning of the chapter, we can affirm that each one of us makes a simulation of what is outside our brain - if there is anything at all - and to make it even more interesting, each one of us creates our own simulation, our own interpretation of the world, giving more importance to certain things, elements, nuances, to the detriment of others, which in turn can play the main role in the interpretations of other people (or entities).

2. Limitation of sleep hours.

We all have dreams when we sleep. Even some other non-human mammals seem to have dreams and, in their dreams, they move and you see that for example a dog is running.

Another issue is if we remember our dreams. Some people remember their dreams without much effort. Others, like me, barely remember anything. There are techniques to try to remember dreams. I have tried some of them and they really work, so I am afraid that, like many things in our lives, it is a matter of perseverance.

And why should I care if I remember my dreams or not? We are supposed to need to rest and from time to time perform a little reset in our brain, and that is why we need to sleep. Likewise, the sleeping brain is not supposed to stop working at all and it flows and wanders from one place to another and maybe those thoughts connected or not are the dreams.

But going back to the previous question and trying to answer it I will tell you the following: according to the general rule it is said that the human being needs about 8 hours of sleep a day. Obviously, this would be the average with many different variations up and down. But let's take this figure as a good one, i.e., 8 hours of sleep a day. Let us also remember that the day has 24 hours. If the day were a cake, 8 parts out of 24 would represent 1/3 of the cake. We divide the cake into 3, and 2 parts of it are awake - some more than others - and 1 part is asleep.

I have not discovered anything new here. But let's put these figures in perspective. If 1/3 of the day we are asleep, that means that 1/3 of our lives we are asleep. That means that, at 10 years of age, I will have been asleep 3.3 years, probably more because babies and children sleep more hours, but let's take the 1/3 rule as a good one. At age 20, 6.6 years asleep. At age 30, 10 years asleep, at age 40, 13.3 years asleep, at age 50, 16.6 years asleep, at age 60, 20 years asleep, at age 70, 23.3 years asleep, and so on. I have the impression that, seen in this way, we have a point of view that perhaps not all of us have been able to perceive.

At age 30, 10 years asleep. At 60, 20 years of our life that we have been asleep! It seems like we are being robbed of part of our life, or that we are using this limited resource very poorly.

Now, what can we do? I can think of two options to at least try to use all those years of our life in a more profitable way.

The first option is very simple, the second is more complex.

The first would be limited to changing our sleeping habits, basically trying to sleep less hours. And here I am not going to get into the problem of whether that is healthy or not. There are people who need more sleep and there are people who need less sleep. But if there are people who function perfectly well on 5 hours of sleep, why don't we all try to sleep a little less? If you usually sleep 8 hours, sleep one hour less starting today. I consider that 7 hours a day to sleep is more than enough.

But what does that hour mean? It means an hour that you have that you did not have before, or at least it seems that you were not taking advantage of it. One hour of savings in a year is 365 hours.

Taking as a reference value 8 hours a day asleep, that is, 1/3, a year we would also have 1/3, that is, in a year we are asleep 365/3, 121.67 days. Now if we sleep one hour less per day during a year, that is 365 hours that if we convert it to days we would have:

$$365 \frac{hours}{24 \ hours/day} = 15.2 \ days$$

Let's repeat, 15.2 days per year. From the 121.67 days a year that we were asleep, reducing the hours of the day from 8 to 7 hours asleep, we are now at 106.46 days asleep.

That's right, by cutting one hour of sleep, we have 15.2 days a year to do as we please. That's more than two weeks. It's as if we've suddenly been gifted two extra weeks of vacation.

We can read, write, play an instrument, ride a bike, go for a run, whatever comes to mind. It is better to dedicate that extra hour to what we decide, and not to enslave it in advancing with pending issues. Let's be clear, whatever we do, there will always be "pending", so let's not use our extra hour in this.

Apart from this, if we start with this routine, we will begin every day with an activity that we like, so we have a good reason to get up, knowing that we will enjoy our first hour, and then with our spirits up we can face the day in a much more optimistic way, although we may have to drink a little stronger coffee.

We can also cut back our sleep hours even more, although then we will probably need to take one or two short naps during the day when we find that we are already more asleep than awake. Ten to fifteen minutes is enough to get rid of the drowsiness of sleep. When we see that we cannot sleep any longer, we lie down and set our alarm clock to go on after ten or fifteen minutes. If we sleep more than those ten or fifteen minutes, it is almost certain that we will wake up completely dazed when we manage to get up, first, and second, then at night, it will be difficult for us to fall asleep.

Obviously to use this technique we have to be free to take that little nap when necessary, and this is not always possible either because of the work or the environment. But if you want to, you can. In all jobs there are breaks and dead times in some way, and it is always very positive to socialize, but let's use one of those small breaks for a nap. Where? Find a quiet room, go to the car, earplugs in the ears and eye mask in the eyes and practically anywhere, obviously indicating to your coworkers what you are trying to do, and even more, try to infect them and make them do the same.

Sleep habits can be taken even further. Some time ago I read that Da Vinci slept in blocks of four hours awake and twenty minutes asleep. This is called polyphasic sleep, which replaces the usual monophasic sleep model by replacing this monophasic sleep with several naps divided throughout the day, in an effort to gain more time and increase productivity. And so, the legend of Da Vinci tells of his great productivity. He would take a twenty-minute nap and get up to paint, which he would do for the next four hours. He would continue with another short nap of twenty minutes to get up as a painter again, or why not as a sculptor, inventor, anatomist, architect, paleontologist, botanist, scientist, writer, philosopher, engineer, musician, poet, town planner, ...?

This polyphasic sleep is already something extreme and not for everyone. Honestly, the approximately two hours that, according to legend, Da Vinci slept do not seem sufficient at first sight, but I repeat that the eight hours also seem to me to be an excess.

I think the simplest thing to do, if you feel like following my advice, is to start with a normal night's sleep of six hours and fifteen minutes (if you find it excessive, start with seven) and a scheduled nap of twenty minutes in the middle of the awake phase, which, as various cultures have shown us throughout history, usually coincides with the period of time following the meal.

And this would be my first option to somehow take advantage of the one third of our lives that we usually spend asleep. In short, try to sleep less.

3. Controlled dreams

The second option we would have, and the one I am thinking of to take advantage of the hours we spend asleep, and which is somehow related to what was discussed in the chapter on living in a simulation, would be that of the dream world.

During a "typical" eight-hour night we go through four or five complete cycles of what is known as the sleep cycle, which may consist of the following phases:

- Numbness;

- Light sleep;

- Transition;

- Deep sleep; and

- REM sleep.

In this last phase is when we are supposed to have dreams, and if we go through the cycle four or five times, that means that on average if we sleep eight hours, we should have four or five dreams every night. However, very few people are able to remember these dreams.

My idea is to encourage to remember dreams, because in the end it is something that is part of our life, and going even further, to control dreams, which, if at first glance seems impossible, it is not.

Once in the dream, if we are able to realize that we are dreaming and do not wake up, we should be able to master it, which opens up a whole new world subject to the whims of our fantasy. Do we want to fly? Let's fly. Do we want to go to other planets? Let's do it.

Later I will also review what the combination of the two ideas above entails, i.e., reducing the hours of sleep and controlling or at least remembering dreams.

But now, let us return to the world of dreams. It is indeed normal that we do not remember our four or five dreams per night, but from time to time, we do remember some dream in greater or lesser detail. If we remember one dream from time to time, that should tell us that we should be able to remember all of them. If we remember our dreams, that tells us that the number of hours we are asleep are not really wasted, because I have been dreaming and my dreams are what have led me to be the king in the underwater world of the water elves or in any other story I can think of in my dreams.

In addition, we know that the time and even the timeline within dreams is not the same as the duration of time and its line when we are awake. If a complete sleep cycle, from its first stage to REM sleep lasts between 90 and 110 minutes, and we only go to the REM stage where the dream is with about 40 or 50 minutes, the rest of the dream could still be considered as unused time, but if during only one dream, within the dream, our experience is hours or even years, the gain is huge.

Therefore, we have and must succeed in remembering dreams. That would already be a considerable gain. If we add to that the possibility of being able to control those dreams, the gains are multiplied.

Many times, we find ourselves at a crossroads from which we do not know how to get out. Or maybe we are looking for the solution to a problem. And the more we think about it, it seems that the further away we are from the solution. However, when we leave some distance and/or space between us and the thing that worries us, the solution presents itself. These are the "eureka" moments, which are explained by the fact that, although we consciously no longer think about it, our subconscious is still working on the problem. Many of these eureka moments occur when we are walking, or simply when we are relaxed and let our thoughts wander where they will. But also, many of those eureka moments occur in dreams, and sometimes we are able to remember them, but other times, no, and we perceive that we know the solution, that we have it on the tip of our tongue, but we are unable to define it.

Now let's imagine that we are able to always remember our dreams. We would solve the problem when the answer is revealed and comes in dreams.

Let's go one step further and say that we are able to master our dreams, and within our dream, we remember what is troubling us in our waking life. Even if we are able to be aware that we are dreaming, it is more than likely that we are also able to go into the dream with the issue present from instant zero with the intention of working on that issue in the dream.

So, we decide to discuss our problem with our childhood friends who come up in our dream when we think of them, or with our ancestors, or why not with Churchill, or Lincoln, or Einstein, or Seneca? Or better yet why not all of them gathered in a big room where I expose, and then each one of them begins to give opinions?

In the end, a dream is a simulation in our brain. A simulation in turn of the "reality" in which we believe or that we perceive when we are awake, but with a few very relevant nuances. First: neither time, nor space, nor in general any of the physical laws are immovable, i.e., they can be modified according to the interests involved in the dream. Second: if we manage to realize that we are dreaming without waking up, if we want, we can be the masters of the world, the Universe and the beyond, making everything and everyone obey our whims, including space, time and the laws of nature.

With these possibilities before us, how can we say no? At least why not try? If nothing comes of it, we lose nothing. We would basically stay as we are now. But if we manage to remember the dreams, we would already have a significant enrichment. And if any of those dreams are controlled, as they say in Toy Story, "to infinity and beyond".

My personal experience is that when I was a child, I remembered more dreams, but after a certain point, I stopped remembering practically all my dreams. It wasn't until recently that I became interested in the world of dreams, and by reading a few books on how to remember and control dreams, I finally managed to remember a dream or two.

Applying the recommended techniques, which by the way are nothing to write home about, I have managed to remember sometimes a lot of days in a row, one, two or even three of my dreams in one night.

Not being an expert, far from it, I believe that the basic technique is self-persuasion, the inner conviction that tonight we are going to dream, and that we are going to remember. During our waking day we should randomly tell ourselves and ask ourselves if we are in a dream, or tell ourselves directly that we are in a dream. In this way, this habit becomes a habit that we even repeat when we are asleep, and then the magic begins.

Sometimes from my dreams I remember a whole plot, and sometimes only parts of it. As if I am watching a movie, but where I have to constantly get up to go to the toilet, to the kitchen, and I am only able to see parts of the movie, but not all of it and I am missing the main plot of the movie.

Other times, days and weeks go by without me being able to remember anything, it is then when I become aware of it and try to concentrate again to remember my dreams.

At other times, I know I've dreamed, but I don't know what about, and it's quite frustrating.

All the books I have read on the subject recommend having a notebook and pen to write down what you remember from dreams as soon as you are aware of it, which can be between sleep cycles in the middle of the night.

I also believe that it is a positive practice and that it stimulates us to remember dreams, or at least to think about that possibility when we get into bed and see that notebook and pen ready on our bedside table.

Like everything else, it is not easy at first and requires practice and perseverance.

Even I have been able to approach the world of dream control on two occasions. On the first occasion I was in the bathroom at home with my daughter when I don't know why I realized I was dreaming, and to check it, the best thing I could think of was to see if I could fly. So, I tried it, and to my surprise I started to rise above the ground as if I were floating. I was going to open the window to "fly out there", but in the excitement of knowing that I was flying in my dream, I ended up waking up, but with a very clear and vivid image of the experience just lived.

The second time was even shorter. Again, while dreaming I realized it, and I tried to go the same way, that is, to start flying, but again in the excitement, I woke up even before I had "taken off" from the ground.

On other occasions, although these are more in my past, I have suffered dreams-chapters of "immobilization". It happened to me when I was a child or rather adolescent, and sometimes also when I abused alcohol one day in my youth.

In these situations, it seems that your brain has woken up, but your body has not. You are aware of your surroundings, or so your brain thinks, you are in your room, in your bed, everything normal, but your body does not respond, you are unable to move, you cannot move any muscle in your body. Obviously, the first reaction is panic, what's wrong with me? I'm sick, I need help... and you try to scream with your mouth that doesn't even flinch. In panic, your breathing becomes agitated, and your heart rate accelerates, which makes the panic situation worse. It seems that you are going to have a cardiac arrest and that your days are over, that you are going to die in a few more minutes and that the next day someone will say that you have died peacefully and painlessly in your sleep, while you are going through a nightmare...

I have to suppose that from such a situation Franz Kafka himself would get his idea, at least initially, for "The Metamorphosis".

With the passage of time, panic leaves some space for reflection and acceptance: Ok... I can't move, accept it and calm down, you'll get something... Little by little you calm down, you try to control your breathing so that it slows down, as well as your heart rate. You are calm, but you still can't move. Calm yes, but your brain is working overtime trying to come up with an explanation for what is happening. Two possibilities begin to take hold.

The first, as is obvious, is the worst thing you can imagine, that is, you have become completely paralyzed, but no, it cannot be only this, since you cannot even move your eyelids, you are in a more catatonic state, you are in a coma ... Well, there is little you can do, you think that someone will come and at least you can hear what they say and maybe you will know what has happened to you. It is not a very positive outlook, but something is something, and in the meantime, you can try to keep moving some part of your body to see if it responds. And if you go a little further and look on the bright side, you can always say that your brain is still functioning, which is no small thing. As I think Seneca said, you can kill me, but you can't hurt me. You can do what is in your power to kill me, but what is inside me, that is something unattainable for you.

The other possibility that remains is that you are living a dream, albeit a rather strange one. Once this idea occurs to you, and even more, that you are convinced that this is so, everything starts to change, and again two possibilities open up, either to continue in the dream, or to try to wake up.

For both situations you have to be sure that you are dreaming at 120%, since somehow getting out of that situation means controlling the dream.

If we decide to change the dream, we can imagine that our lover enters the room, gently wakes you up and you end up making love...

If we decide to wake up, we start by, again, trying to move something or turn around or something in our dream. As before it will seem impossible, but now knowing that it is a dream, we will succeed and our body is irremissibly turned towards waking life.

Another experience I remember having and contrasting goes back to my teenage years. All of a sudden and you don't know how, you're falling. You're going to kill yourself and you're going to be stamped into the cement like you're a fly. However, I was able to imagine some of those clothes hangers that many people in some countries install in the windows of their apartments in multi-story buildings. As I fall, I try to reach with my hands to some of these hangers, to their strings. I don't succeed in the first attempts, but they slow down my fall a little bit. Finally, I end up holding on to one of these ropes, I move hanging by my arms to the window that "coincidentally" is open, I climb up to the window and get into a room of a house I don't know. In the room I just got into, there is an extremely sexy woman inviting me to share her bed... These dreams in which you start as in a nightmare falling, and end up in a sexual fantasy, were repeated a few times, until the moment in which the nightmares of falling stopped occurring, I understand that because really these nightmares were no longer so much to have a happy and known outcome.

4. Parallelisms

Well, after this introduction to the world of dreams, let's look for parallels or relationships with a possible simulation. It would be something like a dream within a dream. What would be the awake phase and the asleep phase in that simulation? In the simulation, why would the sleeping phase be necessary? What would the dreams in the sleeping phase represent in the simulation?

We have to assume that we are in a simulation. Then, the phases of being asleep and therefore also when we are dreaming, I understand that they would be like periods of "maintenance" of the system. If you say that we sleep so that our body both physically and mentally recovers, takes strength, makes a small reset. Taking the parallelism to a simulation and the world of computers, it would be something like deleting temporary files, a defragmentation and other routine tasks for better performance. In this field, dreams would be something like loose fragments, thoughts, illusions, fantasies that cross in front of us in their transit to be erased or rearranged in another location.

We could even say that it is a failure of the system, since that should not happen. Are dreams a failure in the simulation system in which we are living?

Obviously, this is all very speculative, but as I said before, assuming that the premises are true, that is, that we are in a simulation, and that dreams could be a weak point of that simulation, a small bug in the program, this should encourage us to become hackers to take advantage of that bug in the program and try to get as much as we can, to live our fantasies in our dreams or better yet, to try to discover something "real" within the simulation.

In other words, this would be one more reason to investigate our dreams, remember them and try to master them.

5. Let's quantify

Let us return for a moment to how we perceive the world, and the interpretation we give of it.

In theory, when we perceive, we are somehow making a measurement. And by the very fact of measuring, we are interacting with the system, and, therefore, we become ourselves parts of the system we are trying to measure, so we finally distort the measurement.

Let's take a closer look at all this, since the above mentioned also has some interesting links with quantum mechanics and even with Buddhism, by including the meter in the measurement and the disappearance of individualities since everything is part of the whole.

Quantum mechanics is a purely practical theory (or hypothesis...). That is, by applying its extraordinary set of rules and mathematical calculations, we obtain results that coincide with reality.

However, it leaves much to be desired in terms of giving us an interpretation from a philosophical and even physical point of view that satisfies us.

I understand perfectly well what Heisenberg's uncertainty principle expresses, but the fact that I understand it does not mean that I am satisfied with it. This principle dictates that it is impossible for certain pairs of observable and complementary physical quantities to be known with arbitrary precision. We are off to a bad start: "impossibility" ... We have already padlocked some doors....

For example, it is impossible to know simultaneously the position and velocity of the electron, or any other particle, and, therefore, it is impossible to determine its trajectory. The greater the accuracy with which you know the position, the greater the error in the velocity, and vice versa.

That is, if at a given moment, I know perfectly well the position of an electron, I will have an absolute ignorance of its velocity, and, therefore, a second later, I will have no idea where it might be. Likewise, if I know its velocity precisely, I do not know where it is with a tremendous error in its location.

The deal here is to reach a compromise between the two parameters involved in order to know more or less with some certainty something about both. Pretty sad, isn't it?

Basically, it is entering the world of probabilities. In the quantum world, everything is defined by a formula called a wave function, which is a formula whose result is uncertain and the most it can provide is a probability that something will be fulfilled when it has been given the basic data to feed that formula.

Let us now look at the example of Schrodinger's cat to see how far we can degenerate the basic concepts of quantum mechanics.

We have a system consisting of a closed, opaque box containing a cat inside, a bottle of poison gas and a device, which contains a single radioactive particle with a 50% probability of disintegrating in a given time, so that if the particle disintegrates, the poison is released and the cat dies.

At the end of the set time, the probability that the device has been activated and the cat is dead is 50%, and the probability that the device has not been activated and the cat is alive has the same value, i.e., 50%. According to the principles of quantum mechanics, the correct description of the particle at this moment defined by a wave function, will be the result of the superposition of the disintegrated particle - non-disintegrated particle, which more or less we could give it a validity since it does not imply too much in our daily life. But that particle is inside a system with a cat, and the superposition of state of the particle, implies the superposition of states of the whole system, including the cat, with a superposition defined by its wave function of dead cat - live cat. And so, the system is defined. However, the cat cannot be both dead and alive at the same time, but we have no way of knowing which state it is in.

Now we open the box and discover that the cat is alive, or dead, and the mystery disappears, but we have introduced a new element: ourselves. We have opened the box and checked how things are, which is the same as saying that we have taken a measurement.

Where does this take us? The cat was peacefully in its supernatural live-dead state and we have opened the box and the happy, supernatural live-dead cat, we have turned it into a happy live cat or a not-so-happy dead cat. We, with our super power of opening boxes, are able to change the state of the cat. We are the ones who through our measurements define the world as it is. If it weren't for us going around measuring, observing, listening, opening boxes, etc., the world would be an accumulation of probabilities with nothing defined. Thank goodness we are here with our superpowers! This path is absurd, we go back to the Middle Ages and we keep saying that the Earth is the center of the Universe.

dAnother way out? We are part of the system. By measuring, we expand the environment of the system and we ourselves are part of that system, and as Newton said, where there is action, there is reaction. We measure and distort the measurement with our coarse systems, we are unable to see without touching, our fingers are too fat to use this mobile. We distort the world with our actions and measurements and at the same time we are influenced at a physical and psychic level by interpreting what we believe we have measured in an aseptic, surgical, scientific and professional way, when in reality it is as if we were frightened elephants that have just entered at full speed into a glass shop.

We also cannot let this opportunity pass without mentioning Young's experiment, better known as the double-slit experiment. Photons, or particles of matter, produce a wave pattern when they are passed through two slits. We have an emission source, which is thrown against a screen with two slits. Beyond that, a screen where the experiment ends. If what I throw are balls, they either hit the first screen, or pass through one of the two slits, to end up on the second screen. If what I emit is a wave, like a small wave in a container of water, that wave, part will bounce off the first screen, part will pass through the two open slits, and will create an interference pattern on the second screen.

But what is an electron, or a photon, particle or wave? How do they behave in this experiment?

Well, if we launch a single electron that is able to pass to the second screen, we will see a little dot, that is, it seems to have passed like a ball through one of the slits. However, if we repeat this experiment many times, what we get on the second screen is a wave interference pattern, and not the reflection of the two slits.

What is going on here? Do the electrons know what we are doing beforehand to behave like this? Are they passing through the two slits at the same time in wave form taking advantage of all the probabilities given to them by their wave function within quantum mechanics, to then interfere with itself, and reveal itself as a particle at the moment of measurement on the second screen? Does it know that the rest of the particles in the experiment are going to behave the same way?

This is all very strange. Moreover, if we put a little spy behind just one of the slits to find out where one of the particles passes, we find out where it passes, but we destroy the interference pattern, to go to a reflection of the two slits on the second screen. It is like I am an electron and I go to the two slits and if I see that there is nobody, as I am very curious, I pass through both slits unfolding myself, interfering myself posteriori, and before embedding myself in the second screen, I reassemble myself into an entity. However, if I see that there is someone eavesdropping on one of the slits, I am timid to show my super powers of unfolding, and I usually pass through as a particle waving at the spy, who must be very clumsy because I seem to have known about him from the beginning.

All very strange. I have total respect for this theory, tested and proven over and over again with enormous precision, as I have for all its architects, but as I said before, at the conceptual level, it does not satisfy our interests. I am not even saying that this theory is wrong, but it seems to me that, if it is not wrong, it is incomplete.

We can be satisfied by saying that no one has ever told us that the human being is capable of understanding nature completely, but if that is the attitude, we might as well not move forward with anything, because there will come a time when our silly brains will not give more and we will be unable to assimilate what we have in front of us. However, that has never been the attitude of the human being. We may have many individuals who give up in the end, but there is always the occasional stubborn person who perseveres until they get what they want.

Another intriguing possibility is that it is an error in the simulation in which we are involved. Just as I have also commented on it with dreams, I also comment on it here.

Perhaps we are in a very, very good simulation, but in which there are innate errors because perhaps the creators of that simulation did not think that the entities within the simulation were going to get to this point in the investigation of matter.

Perhaps at that level, the simulation is not defined in such detail, and it is still a cloudy cloud.

Or maybe the same entities or things creating the simulation have the same problem of definition at these scales and are using our simulation to see if we discover something that they can then apply to their world. We would be like their experiment where I understand that our time flow is much more accelerated than theirs to see if we are able to discover something.

Perhaps even the entities or things creating our simulation are themselves thinking that they are in a simulation, and if so, how far would the simulations of the simulations go? Would there be something real? Could there be a situation of the whale that eats its own tail? That is, that the daughter, granddaughter, great-granddaughter of a simulation is in turn the mother of the simulation closing in a loop? Would there be something outside that loop?

Another possibility that comes out of quantum mechanics is that of multiple and parallel universes. Since quantum mechanics is basically a probabilistic theory, what the theory of parallel universes says is that, instead of defining a single possibility, all of them are defined, but in parallel universes.

That is, if I can take path A or path B, in "normal" physics, there is a 50% chance that I will take path A, and a 50% chance that I will take path B. In quantum mechanics we would have a wave function that would define these possibilities. If I am the system, once I make up my mind and start walking, the wave function collapses and goes on to define one of the two paths with certainty. If I am only part of the system and the measurement depends on something or someone else, when the measurement of where I am walking is made, again the probabilities disappear and it defines whether I am on path A or B.

However, the theory of parallel universes says that, if there is a possibility that I take path A or path B, what I actually do is to take path A, and path B. How is this possible? Very easy, one universe is created for each possibility that exists.

There would be a me in the path A, and a second me, in the path B. The universe has split in two to admit the available possibilities.

Now, does this mean that the universe keeps multiplying for every probability that exists? At the level of people? animals? climatology? elementary particles??

According to this theory, the answer would be yes. But isn't that a brutal waste of... universes? Well, I would say yes. One thing is that there is a mathematical possibility that this theory of parallel universes exists, and they do not stop multiplying exponentially, and quite another thing is that this is a reflection of our reality, whatever it is, or rather, a reflection of our ignorance.

Arguably, either this theory is a story to cover up our ignorance, or, again, we are looking at yet another clue to the simulation. Let's look at it.

Quantum mechanics has been proven time and time again to work. It does. But at a conceptual level it is still a tool to use the world, not to understand it. It is like a hammer, it is good for what it is good for, to drive nails, but don't ask it about the meaning of life.

Faced with this lack of explanation, the theory of parallel universes, mathematically possible but doubtfully true, emerges as a possible alternative. It should not be allowed to explain a problem with a solution that poses so many problems that it leaves us completely helpless and unwilling to ask. What happens with all those universes? Where does so much energy come from? Can they interact with each other? Is it possible to travel from one universe to another? Is it possible to meet a double of yours? Can one universe eat another? Great... It's like when it was said that the Earth was on top of the shell of a giant turtle that goes around the Sun at its leisurely pace. And where did that turtle come from? And where does it walk?...

It seems to me, therefore, an explanation, yes, but unreal, that the only thing it does is to hide our ignorance, and we go out into the arena with this pompous explanation as if it were a great discovery.

Now, if we look at this whole movie of quantum mechanics with its probabilities, and even with its derived interpretation of parallel universes within the framework of simulation, everything becomes cleaner.

Let's imagine this simulation, our simulation, perhaps our reality, as a computer adventure game. We, the protagonists, start the game, which, like all games, consists of overcoming obstacles by making certain decisions and overcoming levels. The game is prepared to admit several alternatives (not infinite), depending on the decisions we make. Let's see it with an example.

Suppose we are an explorer. The game starts and we start walking. In fact, we do nothing to make our character walk, but he walks by himself. It would be like the passage of time in our reality, we do nothing, but time does not stop running. Tick tock, tick tock, the tireless clock of time. At a given moment in the game, we have to make a decision whether to go right or left. If we don't make any decision, we go to the center where the road ends, and so does our time: game over. Let's take the left and the game continues. But we could also have taken the right, and we would also have continued the game, albeit in another alternative variant of that game. That is, the game is prepared for those two possibilities, and even if we have only used one of them, the game was prepared for both. Just like our reality. Even if we always end up taking one path, our universe-reality is prepared for any alternative, that's what quantum mechanics tells us. Whether I take the left or right path, is irrelevant, the game admits any of these variants, but that does not mean, if I took left, a parallel universe is created where in my game I take the right... That is availability, probabilistic availability if you will, but not reality.

And now tell me, isn't this game of our intrepid adventurer a simulation? The answer is undoubtedly yes. It is a simulation, an admittedly crude simulation with respect to what we imagine as our simulation reality. Could our universe therefore be a really elaborate simulation that is prepared to admit any possible alternative? My understanding is that it could. Moreover, it could even be that it is not necessary for such a simulation to be really "so" elaborate, and that it only admits the possibilities, alternatives for which we are "programmed".

Here we enter a different but related field: free will.

6. Free will.

Does free will exist as such?

In principle, it seems to us yes, because our independence of acts seems to us to be something fundamental and that only depends on our judgment, sometimes more correct than other times.

But is this true, and don't our genetic background and surrounding conditions undoubtedly dictate our actions? Let's take a look.

We are all born with a series of characteristics that are the inheritance of our ancestors. Some issues are more obvious than others. The physical physiognomy, the color of our eyes, the color of our hair, the length of our fingers, our height, the tendency to go bald, our body hair, our visual acuity and its perfection or the need for glasses... Assuming that the pregnancy has been normal and without complications, we appear in this world with a series of inherited characteristics that seem immovable. No matter how much I want it, I cannot be a redhead if I am brown, at least not naturally. If I am physically male, no matter how much I want it and without surgical intervention, I cannot physically be a woman.

With the exception of these traits, everything else, in theory, is more or less modifiable based on perseverance and hard work, or at least, that is my opinion.

Why do some people seem to achieve more while others seem to go through life unnoticed? Why are some people astronauts, while others are left doing public services, such as garbage collection, which, although they have my respect because they are necessary, is not considered a prestigious profession like that of an astronaut?

Perseverance and hard work. Geniuses as such do not exist. Their results are always a direct consequence of their personal efforts. Mozart, the great musical genius, the child prodigy, was not so great. A child is born, his father is a musician, his older sister is learning music. His boundary conditions are very definite and limited. The little boy learns to play the piano, because that's what he has. From the age of three until the age of ten, it is said that he composes, but really his compositions of this period are mediocre and it is noticeable that they are copies of other compositions. You have to realize that his father exhibited him like a monkey, and I am sure that Mozart's childhood was not a happy one. Nowadays there are many parents trying and forcing their children to become professional soccer players. The game ceases to be a game and becomes a slavery for the child.

Let us now look at Einstein. His spatial relativity if it had not been him, it would have been somebody else. At that time, it was a matter of time and the scientific atmosphere was right. Einstein had the right training for it, and in his work at the patent office, instead of letting the hours go by, he forced himself to think about that physical problem, it's not something that thanks to his genius he came up with just like that, but to his work. And when he went further with general relativity he had to study and work hard with mathematics, relying on mathematicians better than him in that field. While it is true that there is an important conceptual leap shown by Einstein, it cannot be said that it was something sudden thanks to Einstein's genius, but rather, to adequate boundary conditions and to hard and persevering work.

Leonardo Da Vinci. Perhaps of these three characters, he is the clearest example of a persevering worker, constant to the point of obstinacy, tireless and meticulous. His final works are the result of dozens and hundreds of previous sketches where little by little he approached the desired result or the closest to perfection. Repeat, repeat and repeat ad nauseam to achieve the final result. And if you also have a small army of assistants at your service, what more could you ask for?

You yourself can take from our history any character that catches your attention for the merits achieved and you will see that they are the result of hard work and favorable conditions.

With this I do not want to (nor can I) take away merits to our "geniuses", but like everything else, we have to put them in perspective, since they were "just" people, just like you or me.

Therefore, given the right boundary conditions and with hard work and perseverance, anyone can be a "genius" in that environment bounded by those first boundary conditions.

That is to say, if Mozart's father had not been a musician, but a farmer, for example, it is quite complicated to see the little 3-year-old Mozart trying to play a non-existent piano in his father's house. Although perhaps because his father forced him and because Mozart was a hard worker since he was a child, at the age of five he could have been in charge of tilling part of the land with the help of his older sister... And someone would have enjoyed the fruits of that field, but not the music for which he is finally remembered by all of us today. That is an outline condition.

It's like cooking. We are going to make a Spanish omelet. It is very difficult for us to make a Spanish omelet if we do not have the necessary ingredients: potatoes, oil, eggs, salt, a frying pan, fire, ... The ingredients would be the outline conditions. Without ingredients, there is no omelet. Without boundary conditions, there are no results in that field. If it is the first time you make this omelet, even if it is mediocre, out of pride and because it has turned out well, you will be very happy, but when you have made a hundred or a thousand, the omelet will be much better, and probably the execution time will have been shorter. Practice, repetition and hard work bear fruit, leading our results towards "perfection".

Thus, hard work is the ingredient that differentiates our actions between mediocrity and genius, always within given boundary conditions.

7. Effectiveness.

Having said what was said in the previous chapter, here we would have to see what is the difference or turning point for a given person to persevere and work hard, or, on the contrary, to be carried away by life to the grave. We are talking about effectiveness.

All of us by genetics can perform at peak performance as long as the surrounding conditions are right for maximum effectiveness. However, we seem to be lazy and constantly pass up opportunities to achieve that maximum effectiveness, as if it doesn't matter, but that is absurd. We can ask anyone and they will always tell us that they want the maximum, the best, and yet we don't see how efforts are made to that end. Are we in one of those quantum paths, parallel universes and/or simulations where we have been cast in a mediocre role? Who is playing against us? Can we really do something about it and change it?

Many times, people talk about getting out of the comfort zone. That basically means doing things different from what you are used to and what you feel good about in order to get new results. We are trying to change the boundary conditions to see and/or discover our potential.

We are programmed to be great, but we have been let loose in the game of life without the precise instructions to know where we can be great. When someone succeeds, they are lucky enough to be born right where they are predestined to be great, they become important and relevant. The rest go through life looking for their place, or resigned to live their mediocre life in the wrong place.

8. Buddhism.

Let us now turn to a completely different subject: Buddhism. And why Buddhism and not any other religion? Someone might ask, and the answer is obvious: because I am the one writing these lines. And without being so obvious, because I think it is one of the few religious cults that directs you and forces you to know yourself and to look at the world through new eyes and a new vision. Christianity, for example, and I only mention this religion since by birth it is the one that geographically "touched" me, seems to say that everything that happens in the world, and to you in particular, is God's design, and that, even if you do not understand it, everything has its meaning, or even can be a test to see how strong your Christian faith is. Without despising this religion and all its followers at all, given that it proclaims love of neighbor and a set of rules for a peaceful life in society, it is still a brutal act of faith to believe that everything is designed by God and that basically our choices are to be with him, Obviously this can lead to fanaticism as we have already seen happened in the Middle Ages in Europe, or to the current terrorist acts of other religious groups that take over and misinterpret their own sacred texts.

But back to Buddhism. Without being an expert, which I am not in anything, much less in Buddhism, here it is about becoming aligned with the Buddha himself through meditation. I don't know if "aligned with Buddha" is a good expression. It is about coming to the ecstasy of understanding and comprehension through meditation. Meditation and according to Buddhism itself, has several degrees of advancement, which are not achieved overnight, and rather requires perseverance and continuity to achieve minimum goals (here we also require perseverance and hard work ...). The first objective is to leave the mind blank, which as simple as it is to say, is extremely complicated. Close your eyes and try to literally make your mind blank, without thoughts. It is normal that in a few seconds our mind is filled with thoughts, memories, dreams, plans, fantasies and countless other ideas.

It's hard to shut up our brain. As I read somewhere, it's like trying to shut up a bunch of monkeys that live in our head. You can do it, but then one monkey comes in with its screams and then all its companions are screaming inside our head. It is difficult and very complicated to shut up all those monkeys. Therefore, the basic technique is to try to fix our thoughts on only one thing, such as, for example, our breathing, which is constant and continuous.

Counting inhalations and expirations of air. We probably won't even get to count ten of these cycles before a monkey appears. That's okay: we are aware of its presence, invite the monkey to leave the way it came, and resume our breath count.

At first, this seems to be repeated endlessly without getting anywhere. However, over time, the monkeys learn to calm down, and that's when we can see beyond it. We begin to learn to see things as they are and not as we think they are. To see with a new prism. To see without preconceived concepts. To see objectively. We see or try to see things without the permanent bondage of our emotions. We see without judging. We discover once again that our senses permanently deceive us, and that, in addition, our brain transforms what we perceive into something different, motivated by cultural and emotional influences. We see that we cannot trust ourselves.

But at the same time, we discover a new world that has always been there and that we did not see. We begin to hear the wind, and the birds singing, the leaves of the trees dancing, even the grass growing, the snail moving, the intensity of the colors reborn in a permanent spring. We see other people and begin to understand them, their motivations, their desires and fears, we even begin to understand our enemies and without actually sympathizing with their cause, we feel compassion for him and wish to help him. In fact, we begin to feel compassion for everything, since we understand that the whole is connected to the whole, and we are part of that whole. As part of the whole, we cannot wish any harm to any other part of that whole, and we feel only compassion and desire to help.

As I have said, within meditation, there are many levels, and to see the above, perseverance is necessary. Personally, I have only managed to glimpse specks of the above, but perhaps that is enough to perceive the rest.

Perhaps other people will never get to see this. Others will go further and see this universal connection with everything, reaching total ecstasy. Still others will feel the benefit of meditation, as concentration and even just to fantasize, but without going much further.

There are cases where this universal connection of the whole with the whole has also been reported, which in turn gives people a kind of superpower to be happier and live life in a more peaceful and tranquil way. These cases are those where one has been close to death, and has come back to tell the tale. Without reaching these cases, there are many cases of people who, when they die, seem to be calm and happy. Obviously, they do not come back and tell us about their experience, and nowadays that happiness and calmness may be the result of the medicines and drugs that are usually administered in the last phases of life...

However, there are people who return from the same death and their behavior in practically all cases is the same. They see a fundamental connection between all things, living and inert beings, past, present and future is a continuum that only we insist on separating. They give importance to what they believe is worthwhile, such as, for example, spending more time with loved ones and trying to become an improved version of themselves every day.

As I see it, when you are about to die, your brain turns on a total emergency red light, and gives the order to prioritize. It's like when you try to start a car and it's running low on battery. The auxiliary and non-essential systems are turned off to prioritize the main function. Lights, radio, ventilation, everything that can consume energy and take it away for the priority function, starting, are turned off.

When we are near death and at the red emergency light, maybe all our crazy brain monkeys shut up and we see everything with crystal clarity.

Could it be then that when we meditate, we "play" at dying?

Both people who have experienced a near-death experience and those who have made significant progress on the ladder of meditation, say quite similar things: they see the world from another perspective, giving more importance to the song of a bird than to the traffic in which can sit for hours, feel a connection with everything and everyone, not wishing evil to anyone and just feeling compassion and desire to help, have an inner peace as if they had discovered the mysteries of the universe and life, ... In short, I would say that there are several similarities. In fact, through meditation one can "accept" death as just another phase in our life, and without major trauma. As I read somewhere: It is normal that we die, but we only do it once.

Obviously, Buddhism pure and simple assumes that death is one more step towards its next life through reincarnation. But let's leave that part of reincarnation out, at least for the moment, because to me, it's another story that sounds like a fable.

Therefore, and following this line of argument, we could say that meditation is a course to know how to die, acquiring in turn the advantages of a vision and perception typical of the pre-death phase, but ahead of time, during years of life, to give in turn to our life a broader sense of our purpose in passing through this world.

9. Connections.

In any case, what interests me about this whole film is the theme of connection; the connection of the whole with the whole, the feeling of being part of a whole, as well as the relief and satisfaction that this gives you, within our hectic and constantly unsatisfied lives looking for more and more.

That universal connection is present in many places, obviously, as it should be, since it is universal connection, and, therefore, it is not that it is in many places, but it must be everywhere. We find this connection in death and in meditation as we have just seen.

We also find this connection in quantum mechanics, or in a broader scenario, in all the physics and chemistry of our universe, since as we saw, we cannot take any particle separately, since no matter how hard we try, it will always be in interaction with its surroundings and by extension with the whole Universe. And in the unlikely event that we managed to completely isolate a particle or a group of them (which we would have to see how, since I do not see it), we ourselves would perform some measurement interacting and thus disturbing the isolation. And as we are in turn constantly and without pause interacting with the gravity of the Earth, even with that of the Moon and the Sun, with the air around us, the water that hydrates and washes us, etc. ... we are back to the starting point where we constantly interact with our surroundings, and by extension with the entire Universe, and connecting again the whole with the whole.

We also perceive this connection of the whole with the whole in the world of dreams, once we interact with them, that is, at the moment we are aware that we are dreaming and take control of the dream itself, where we can go wherever we want, be whoever we imagine and proceed in various ways. How is it possible that all these possibilities are in our head? They are in our head because everything is in our head, or at least the possibility of everything being in our head is present as part of the whole.

Also, and out of context, but I find it interesting to mention, astronauts when they are in space and see the Earth from their exceptional point of view, perceive and physically see that connection, if not total and universal, at least that of all living and inert beings that make up the Earth as a tiny planet around a star called the Sun.

Taking a step further and looking at the possible simulation in which we could be immersed, this connection of the whole with the whole is equally self-explanatory, since everything would be integrated within the same "computer program".

10.   Astronomy.

Now we are going to talk about a different, but related subject 😊, astronomy. From a historical point of view, a lot of progress has been made, but from a more generic point of view, we are in kindergarten, and probably because of us.

The Earth has always revolved around the Sun, but historically until Copernicus, it seems that it was the other way around in the mind of mankind, although physically that was never so. We saw in the 20th century that the Solar System where we live is on the outside of one of the arms of a galaxy, we have called the Milky Way, which we can see with the naked eye on clear nights. This galaxy is one more in the cluster of local galaxies. These clusters create filaments, which are large groupings and finally we contemplate the observable Universe.

The observable Universe is said to be observable because our sight reaches us up to a certain point; beyond that point, we intuit and suppose that the rest of the Universe follows the same patterns and nothing changes. Actually, it is not as far as our sight reaches us, and that has been a simplification, but to explain it better, we must first explain another series of issues.

It has been proven that all galaxies, except for those that are closest to us, are moving away from us. If we assume an egocentric and somewhat ignorant point of view, we will again think that it is because we occupy a privileged place in the Universe. But it takes very little imagination to see that this is not so, since, if we take as a reference a galaxy that is moving away from us, and now imagine that we are physically in it, we would have the same observations, i.e., that all galaxies are moving away. This conclusion has been reached after studying the light spectra of these galaxies, where it can be seen that there is a shift towards the red band, it is the Doppler effect; when a train passes by whistling its note seems sharper when it approaches and deeper when it moves away, due to the shift of the sound wave. To the speed of the sound wave we must add or subtract the speed of the train. Something similar happens with galaxies, and the conclusion is that the Universe is expanding. To the velocity of a galaxy, we must add or subtract the velocity of the expansion of the Universe. This expansion, in the current phase, is such that the velocities of the galaxies are not significant, prevailing this expansion and making everything seem to be moving away.

The typical example is the balloon where we have painted little dots representing the galaxies. If we inflate the balloon, the space between the galaxies expands - this would be a simplification of the expansion of the Universe.

A few questions might immediately arise from the above, as a start:

- Where is it expanding to?

- How fast does it expand?

- Will the day come when we see nothing since everything is out of sight?

- From where does it expand?

- Why does it expand?

- …

There are many questions in this regard and the curious thing is that we are able to answer, even if only partially, some of these questions.

It is assumed that right now we are initiating a phase of acceleration in the expansion. It seems that until now the mass of the Universe and thanks to gravity, maintained this expansion at, let's say, a constant speed, but we must have passed some threshold where gravity can no longer slow down or even maintain a constant rate of expansion, and from now on this expansion will go faster and faster. Which means that effectively in the distant future, and assuming we are still here..., we will look at the sky and there will be nothing.

And what causes the expansion to be accelerating? That is not clear, but if it really is so, there must be some kind of force and/or negative reaction, something like gravity, but instead of attracting, repelling, as when we join two magnets by the same pole. Here comes into play the dark matter and dark energy that we will talk about later.

But let us now look at the whence? That is, if the Universe is a kind of balloon that is inflating, at some earlier time it had to be smaller, and if we take the movie of the history of the Universe back far enough, we will find that everything should be concentrated in a tiny point. This is where scientists talk about the Big Bang theory. Once upon a time there was nothingness, but absolute nothingness, not the quantum nothingness that is full of creation and destruction thanks to the uncertainty principle, the nothing nothingness, where there is not even space and time. And according to certain hypotheses, that nothingness at some point and time, which is wrong, at least to count it that way, since as we have said, there was no frame of reference neither spatial nor temporal, as I said, that nothingness became unstable and from here the Universe arose: space, time and energy that then "degenerated" partially into matter ($E=mc2$). A few fractions of a second later, some instability occurs again and the newborn Universe expands exponentially for a while. This would be the inflationary model that allows us to explain the homogeneity observed in the present Universe, and a number of other issues. In recent times, this inflationary phase is also being questioned.

In short, there is a story with its strengths and weaknesses that more or less explains the Universe, which is no small thing by any stretch of the imagination.

Now, at the initial moment where we have to deal with extreme conditions, and where in theory we should have the quantum mechanical and gravity formulation together, and not each one, on one side, we are a bit blind, not to say completely blind. At this point, it is all very well to say that nothingness was unstable and the Universe arose. But we can also put in the hand of some stupid deity and say that it was a divine act of creation. And I said stupid deity, because once the Universe is created, it loses absolute control over its creation. That is, there is no record or hope that, if the Universe is the work of a God, he can intervene in any way in its evolution and events. Everything that God had to say, he already said at the moment of creation. If he forgot to put something in that divine moment of creation, ah... it feels, it is already late. Of course, we can also imagine a God who, counting on his superpowers, cooked the Universe so finely that even the smallest detail of its evolution, including our ephemeral wandering, is perfectly programmed.

And with this we return again to mention or recall topics already mentioned. The first would be that of free will. If the

Universe is the work of a God who has dictated everything in detail, we are only executing a predefined program, with no possibility that we can do anything out of our pattern. Again, it seems that concepts such as guilt are out of our hands.

Likewise, if everything is planned by an all-powerful God, I wonder about the differences between that all-powerful God who has created a detailed Universe and a supercomputer or an AI artificial intelligence that is simulating our Universe.

At first glance, I do not see many conceptual differences. And just as there are many believers in a God, or several of them depending on the culture and the place where you were born, perhaps I prefer to believe in an IA that is simulating our Universe.

In either case, it seems that I have no choice in the matter either since I have been created and/or simulated in this way and my real options of choice are pure fantasy.

In any case, this whole story of the Universe and its evolution has a lot of flaws that we don't seem to feel like contemplating.

The told story of the Big Bang contemplates matter such as stars, planets, ourselves, but all that we are aware of and see accounts for 4% of the total! To complete this picture, we must include 30% of dark matter and 66% of dark energy... 😨

Of dark matter and dark energy, we know absolutely nothing 😨 😰. Their existence can be inferred from gravitational effects in matter, as well as in anisotropies of the cosmic microwave background. That is, we know nothing about 96% of what constitutes our Universe, and we can only infer in rather indirect ways its existence. "Avoiding" that 96% of the Universe, with the 4% we have left we can conjecture a theory called Big Bang that gives us peace of mind, since it explains "more or less" where we come from... ignoring the 96% of the Universe that we do not know what it is made of.

I think that "maybe" we still have a "little bit" of work to do... or maybe it's just me...

If we add to this the exotic world of quantum mechanics, we begin to see several things. First, we are far from understanding the deep reality of our Universe. Second, it is incredible that, with our limited knowledge, we are able to predict future events, we have working science, and we have evolved technologically.

Now if we add these two points, we begin to see that something is wrong, and perhaps it is not precisely us, but the system, the system in which we are involved. Because at the end of the day, things work, and we continue to evolve from a scientific and technological point of view. And all this is explained with laws and hypotheses that we have managed to extract from nature. However, when we seem to have the answer to something, it suddenly slips through our fingers. We can say that this has always happened throughout the history of science, but I believe that we have reached a place called the end of the road.

11.   The four forces.

According to what we "know", we have 4 forces or interactions: gravity, electromagnetism, strong interaction and weak interaction.

The strong interaction is the quantum chromodynamics responsible for the atomic nuclei and their components not disintegrating.

The weak interaction allows radioactive decays.

These two forces are of very low range and therefore only occur at the atomic level.

Electromagnetism is a force more understandable to us. Negative and positive fields and has a long-distance range, in fact, "infinite". And "miraculously" the negative fields balance with the positive fields at the cosmological level and no one wins. It seems that we have as many negative charges as positive ones, and the balance is zero. Therefore, unless we isolate some part of the Universe where there is a time discrepancy between charges, we do not feel their influence.

Finally, we have gravity, which like electromagnetism has an "infinite" range, but unlike electromagnetism, we do not have a positive and a negative pole counteracting each other, but only one polarity. And although gravity is of the four forces the weakest, having only one polarity, it is a force that by accumulation becomes the force that dominates, at least partially, many, if not all, phases of the evolution of the Cosmos. It is this force that we constantly perceive and that keeps us tied to the Earth. It is a force that we perceive and intuit from birth because we are constantly experiencing it.

But as always, we have some little problems... There are situations in the Universe foreseen by our own science, in which our masterly conception of forces and interactions fails. Let's look at the most relevant ones: black holes, Big Bang theory, and, "accelerated" expansion of the Universe.

12.   Black hole.

A black hole is a finite region of space in whose interior there is a concentration of mass sufficiently high and dense to generate a gravitational field such that no material particle, not even light, can escape from it.

Within general relativity we could say that the curvature of space-time is such that even rays of light remain orbiting without the possibility of escape.

We can make a simile with the Earth, its gravity and the launching of satellites. Let's imagine a cannon on the surface of the Earth, and we fire it. The projectile will fall by the effect of gravity at a certain distance. If we repower that cannon and fire it again, the projectile will fall a little farther. If we get a sufficiently powerful cannon, the projectile will go so far that it will not fall on the surface and it will be in a situation of permanent fall, in other words, it will be in orbit.

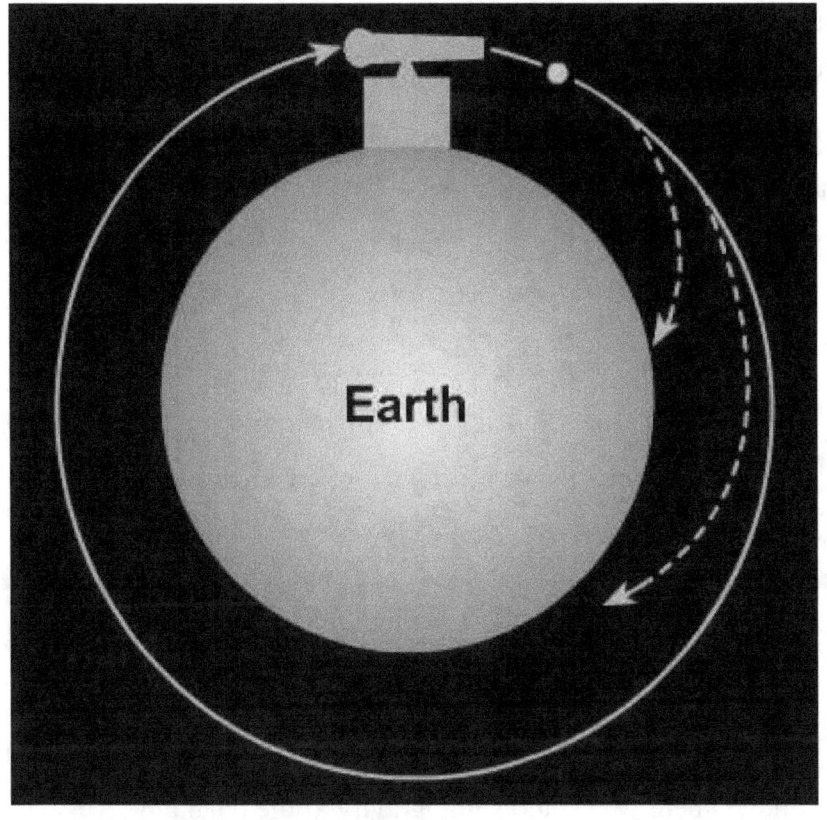

Illustration 1 - Entering orbit.

All satellites and the space station are in orbit, i.e., in free fall. If the velocity is less than what is necessary to maintain this situation, it will eventually fall. On the contrary, if it is somewhat higher, it will be able to escape and will be lost in space.

In a black hole, light traveling at 300,000 km/s is not able to escape, it either stays in orbit or falls (to where?).

But what interests us, once we have seen more or less what it is about, is what is inside a black hole. To create such a space-time curvature, such a brutal gravitational field, the concentration of mass, and therefore also energy (since mass and energy are equivalent $E=mc^2$), must be so high that our laws and our science fail.

The atoms with their electron swarms disappear, leaving the nuclei bare. Even these nuclei come together because the force of gravity is so immense that it overcomes any kind of electromagnetic repulsion. And we go on and on to form a soup of quarks and gluons, entering the world of the very small, quantum chromodynamics. And yet gravity is such that we should go further... Here we have a soup of subatomic particles and all the known forces, but where it seems that gravity still dominates. And it is now where we should study gravity at the quantum level. The only thing is that we don't have a theory of quantum gravity and all attempts to get one have been unsuccessful.

Trying to bring space-time and its curvature responsible for gravity into the quantum world has not been possible to date. A kind of space-time foam has been imagined, but it is not compatible with everything else. That everything else would be the other three forces that do get along with the concepts of quantum mechanics.

So, we have on the one hand quantum mechanics that works well and with astonishing precision and detail. This quantum mechanics from a conceptual point of view is far from being something intuitive and we can only say yes with our head. It works, but it is not clear to us how, its bases, its foundations, the conceptual building.

On the other hand, we have general relativity which also works well and with spectacular precision. Unlike quantum mechanics, general relativity is conceptually "understandable". Not that it is a clearly intuitive thing to see four-dimensional space-time bending by the presence of mass and/or energy to produce gravity, but with a little imagination, it is an understandable concept and one that our brains can assimilate.

However, if we try to put quantum mechanics and general relativity together, we fail. Why? What are the reasons? Let's look at some possibilities:

- Nobody has said that these two theories have to come together... They can coexist and each explain their own. We would still have the problem of knowing what happens in a black hole (and at the beginning of the Big Bang...), but the problem is limited to our impossibility to get into one and see what each of the theories explains;

- One or both theories are incomplete. In this sense, at first, we would say that quantum mechanics is not complete, or is not the final stage of this theory, because, even if it works, there is no one who understands it. But it could also be the other way around, that general relativity is incomplete because it does not foresee and accommodate the rest of interactions;

- Our brain is incapable of understanding the Universe. This could well be the case, but the way our brains are made, this possibility will not stop us in our search for the "truth" and we will continue to speculate, rehearse and draw scenarios;

- We are in a simulation. The simulation is "almost" perfect, but almost. It leaves loose ends like the present for some reason or for no particular reason:

    - The simulation would consume too many resources to reach that level;

    - The simulation, or those who simulate, also have no explanation and are waiting for us to make something clear;

    - The simulation has deliberately left these "errors", empty, meaningless, unexplained. The goal of the simulation would then be something perverse or something like a game of chess. We would be something like television for whatever is simulating. And obviously it has to leave issues unexplained to see how we break our heads and make the show more interesting.

It could well be that these "gaps" would be clues for us to discover without a doubt and in some way, that we are in a simulation and that we wake up from the dream...

And what would happen if we were to discover without a doubt that we are in a simulation? Would discovering and explaining these mysteries of the Universe, and supposing that we are in a simulation, enable us to contact the entity that is simulating us? Assuming that we are in a simulation, would it not be excessively reckless to contact the simulator? I am referring here to the fact that, in a contact with the simulating entity, we could disappoint, piss off, stress, or whatever, this entity, causing it to unplug us, to turn us off, game over. Or merely by discovering that we are in a simulation, the very meaning of this simulation loses meaning and automatically causes its destruction?

And what would we say to the creator entity? It is as if we managed to talk to a God, but knowing that he listens to us and knowing that he has absolute power over us.

And what would be the psychological consequences for us? To know that we are a plaything of something superior?

And now let us note that, if we replace "simulation" in the previous paragraphs by "religion" in any of its variations, we could continue to ask the same questions. Where does this exposition lead us? We are seeing parallels between Gods and simulations. However, trying to explain a simulation seems to us something more attainable, than explaining what, if anything, can pass through the brains of the Gods.

There are also another series of nuances. Believers, in any of their versions, seem to worship and respect their creator Gods. However, if any of these believers were confirmed to be in a simulation, worship would turn into considerable annoyance: what's going on, are you kidding me, am I just a toy/entertainment?

13.   Big Bang.

But we've gotten a little bit sidetracked and ahead of ourselves...
Let's go back to inconsistencies. We had seen the subject of black
holes. Something quite similar happens with the first phases of the
Universe, the well-known Big Bang, if it really happened.

Right now, we see an immense and, within its irregularities,
homogeneous Universe. But it was discovered a few dozen years
ago that, except for the nearest galaxies, all the others are moving
away from us. This is deduced from the red shift of the galaxy
spectra. What happens? Do we have an infection and they all move
away from us? This would be the egocentric point of view.
Opening the perspective, a little, the correct explanation is deduced;
that the Universe is expanding.

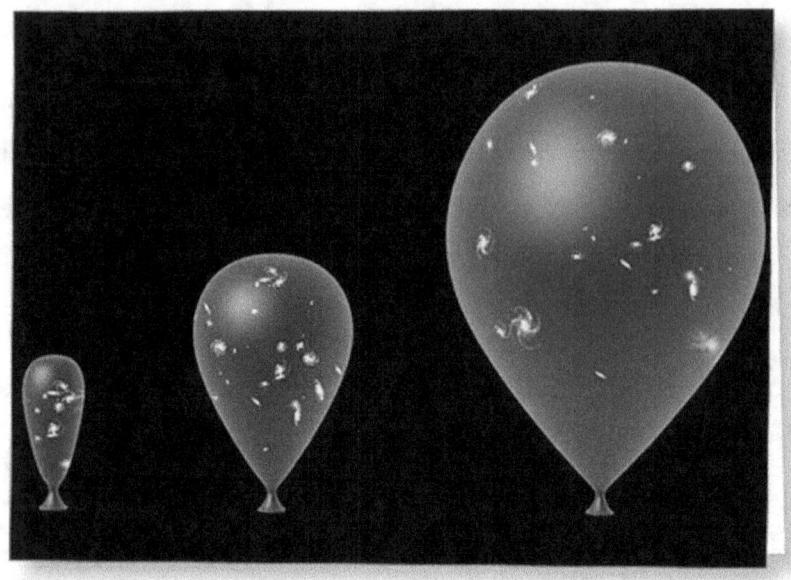

Illustration 2 - Distance increase between galaxies.

Let's suppose that space-time is represented by the rubber of a balloon where we draw some dots representing galaxies. As we inflate the balloon, it grows, as does the space between the dots. If we now take one of those dots as a reference point, as if we were in it, we would see that all the other dots move away from us, and they move away from us faster the farther away those dots are. This is exactly what we see in our Universe (with the exception of the nearest galaxies, in which gravitation counteracts this expansion of the Universe for the moment).

Now, if instead of moving forward in time, we go backwards, we see that the balloon, instead of expanding, contracts. But how far? To a point where all space-time and matter-energy would be concentrated. The problem is similar to that of black holes. Our knowledge leads us to a point where things go wrong and we are left without explanation.

By playing the movie backwards, we are going into the past of the Universe, to its own birth. In those first moments with everything so concentrated, the calculations are made with huge numbers for everything, trying also to combine and involve the four interactions. And of course, if we have not been able to explain this problem with black holes, we are not able to do it on a larger scale with the egg of the Universe.

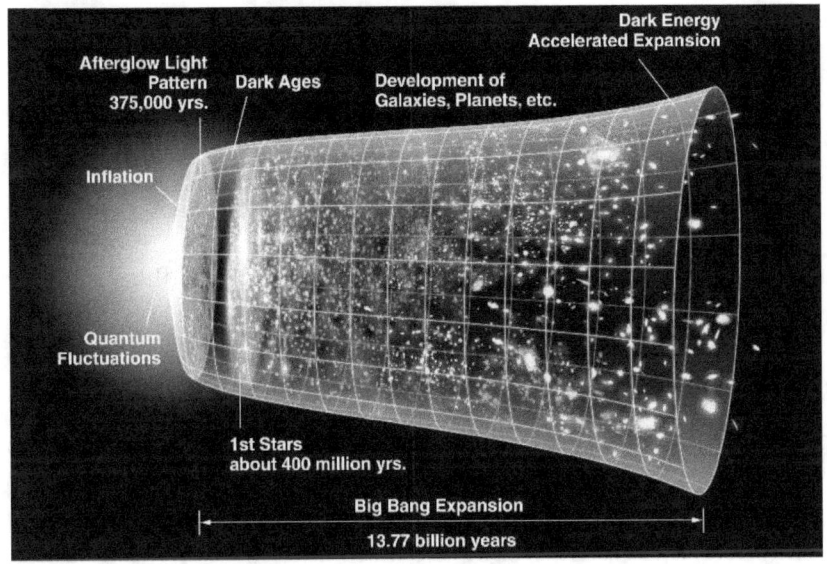

Illustration 3 - Expansion of the Universe.

Now, through our science we can give explanations with enough certainty starting from a few thousand years of the

Universe, and explain more or less (obviating for example dark matter and dark energy...) the rest of its history up to the present moment.

We also have hypotheses and ideas that go back to the very beginning of the Universe, but we could say that the further we go back, the more our degree of certainty decreases proportionally and even exponentially, allowing uncertainty to enter.

So, we have stories for when the Universe was a few minutes old, or even when it was a tiny fraction of a second old (10^{-43} seconds known as Planck time). And beyond that we can even give a somewhat mysterious explanation saying that the Universe started, because nothingness was unstable, but that's too much speculation for my taste...

Another problem that arises with the expansion of the Universe would be the following. If I throw a ball upwards, at first it will rise fast, but due to the Earth's gravity, it will gradually slow down until it finally stops and falls back to the ground (here I am ignoring air friction, wind and other possible interactions in favor of simplicity). If we now compare that ball with the Universe in a rather crude resemblance, we can say that the Universe was born in the Big Bang, that it is expanding, that it will stop sooner or later due to the gravitational effects of ordinary matter, energy, dark matter and dark energy, and that will be the moment when the Universe will start to contract.

Shrink how far? To a single point opposite to the Big Bang, the Big Crunch. What happens at that point is something we don't know since it is again a case similar to that of black holes or the Big Bang. It may disappear, it may bounce, ...

It could also be that after the Big Bang, the expansion slows down, but never stops, because gravity is not strong enough to slow down the expansion and reverse the direction of motion. The Universe would expand constantly.

However, and according to the latest observations, none of this is so. Everything was going more or less as I have commented, that is, Big Bang, expansion and little by little it slows down, until now. This now, which is precisely where we are here, must be understood with perspective, i.e., it is a short period in the cosmological scale, but it is probably our entire history and something more. As I was saying, now and according to the latest observations, the Universe is expanding, and that expansion is accelerating. That is, the whole Universe is growing faster and faster and all parts of the Universe are moving away from each other faster and faster. Everything we see in the sky today will eventually disappear behind what we call the observable Universe (i.e., the Universe we can see because light has had time to reach us). Even our galaxy will dismember little by little and everything will be consumed until there will be nothing left but space with a few scattered particles in a Universe larger and lonelier than ever. That seems to be the sad end of our Universe, and obviously our end, or the end of what we left, as long as if we do not kill ourselves, and in turn are able to leave the Solar System and find other habitats before the Sun burns out, in fact, much sooner, since when the Sun becomes a red giant, it will probably engulf our planet Earth.

There seems to be no turning back, entropy supported by this accelerated expansion will dissolve the Universe. And although this pathetic end will take longer to come than we imagine, that does not mean it is not there. It is just like our own death; we always think we have time left to realize our dreams, until we see that there is no more time. Just because we deny death does not mean that sooner or later it will approach us with its scythe and take our lives away.

Now, what is the explanation we can give to an accelerating expanding Universe? Well, basically none with a sufficient degree of certainty. And here we have one more problem to add to our list of problems. Recall:

- Quantum mechanics;

- Relationship between quantum mechanics and gravity (affecting the explanation for black holes, Big Bang, ...);

- Dark matter and dark energy; and now,

- Accelerated expansion of the Universe.

It may well be that these four problems that I am presenting here have a common root and all four can be solved in one fell swoop. But it may also be that this is not the case and will remain beyond our understanding forever. In any case, and because that is our nature, we will continue to investigate.

Although I do not see any convincing explanation for this accelerated expansion of the Universe, let's look at some of them.

Special zone. It could be that the Universe as a whole does not have this accelerated expansion and that it is only occurring locally, around us. A greater gravity, and therefore accumulation of matter-energy, outside our observable Universe with locally accelerated expansion would cause this. Now, since we do not perceive a dominant directionality for this accelerated expansion, that would mean, to my understanding, having our observable universe with locally accelerated expansion surrounded by a kind of four-dimensional "empty" sphere (space-time) with a higher concentration of matter-energy. But that would make us special... In the observable Universe we do see concentrations of this type, filaments are called and they are concentrations of galaxy clusters, but there is no preferential directionality. Basically, this idea is like saying again that the Earth is in the center, not of the Solar System, but of the Universe due, obviously, because us... An idea that I proceed to discard directly.

Another perhaps more accurate explanation is that since our science is incomplete, we should expect to complete it in order to try to understand something. Within this idea, we have several proposals:

- That gravity at large distances changes sign or at least behaves in a way that allows that acceleration. And this is how physicists reintroduce a parameter in Einstein's equations, which Einstein himself dismissed as his biggest mistake. Although that parameter was introduced for different reasons when thinking about a static Universe, its reintroduction in the equations could give some explanation. But with this we would be back in the deep end. Even if mathematically, it worked - which is still not clear - from a conceptual point of view we would lose the explanation, turning the theory of general gravity with an understandable conceptual explanation, into a second theory like quantum mechanics, which mathematically works, but which is far from satisfying us and calming our soul (whatever it is);

- That dark matter and/or dark energy have something to do with it. And I say, if dark matter and dark energy represent 96% of the Universe... obviously they will have to say a lot about the evolution and future of the Universe... Now, as it is something that we do not handle, nor understand, nor know what it is, it is difficult to introduce influence in our equations. Difficult if we are strict, if we are not so strict, we introduce the elements that we want to make the system work and we blame the dark matter/dark energy for what we do not understand...

  Basically, I personally doubt even the existence of that 96% that does not let itself be seen... Either it is the result of our ignorance or that the system and/or simulation fails;

- A third possibility, and this one is homegrown, has to do with the nature of space. What is space? That frame where all events take place... That stage where all works are performed... Here I am referring to the completely empty space, even of quantum fluctuations with constant creation of particles followed by their destruction without violating any basic principle of nature thanks to the uncertainty principle of quantum mechanics.

One way of looking at space is as if it were an elastic mesh. Is this a simple similarity, or could we go further? What if completely empty space harbored energy? After all, it is assumed that when the Universe was born, space (and also time) took its first steps. Before the Big Bang, there was nothing, not even space or time. We have always seen space as such, i.e., space. Before Einstein, we also only saw matter as matter, but after Einstein, we saw that matter and energy are different sides of the same coin.

Now let's invent a three-sided coin, where we have matter, energy and space-time. Could it be? I don't see why not, and I haven't seen any reference to it. I repeat again: What if empty space has energy? Or further, what if space, or space-time, not only has energy, but is energy in a way we have not seen it before?

14.   The One Element.

As far as we can see, space-time is much more abundant than the matter and energy contained in that space-time, and if that space-time contains some energy (or matter, remember E = mc2), we would eliminate at a stroke the need for dark matter and dark energy. That is, space-time would be the much sought-after dark matter and dark energy, the only thing we have not been able to see, although it was right under our noses. That would mean that the equations of general relativity are incomplete and we should introduce one more element that would somehow make it possible to introduce space-time into the pot of matter and energy, since, according to these ideas, space-time itself should be able to transform itself into matter and/or energy, and vice versa.

Where did this idea come from? Well, basically from the problem of the accelerated expansion of the Universe and combining this problem with some features of quantum chromodynamics, and finally giving space-time a non-zero value of energy, and therefore considering the whole Universe made of a single element, which in most cases is shown to us as space-time, in other cases as energy and in other cases less, as matter.

We are made of matter, and because of our persistence in considering ourselves the most important thing in Creation, it is always difficult for us to remove ourselves from center stage. But as we see now, in this framework of the one element, its appearance and decay as matter could be considered as something residual.

Let us baptize this unique element capable of presenting itself as space-time, energy and/or matter. Let's call it Amale.

The Amale should therefore be able to explain some of the problems we have seen.

We have already said that by endowing space-time with energy, space-time takes the role of dark matter and dark energy, and, therefore, this problem disappears. Space-time endowed with energy would explain the gravitational effects that require its existence, for example, with the rotation of spiral galaxies and their lack of disintegration.

If the disks of spiral galaxies have a mass distribution similar to the observed distribution of stars and gas, the velocities of the rotation curves should decrease over long distances in the same way as occurs in other systems with most of their mass at the center, such as, for example, the Solar System. That is, the farther away from the center, the slower the orbit.

However, this is not what is observed. What is seen is that the velocity curve is maintained. This is explained by assuming the existence of a large amount of dark mass surrounding the galaxies.

With my proposal to endow space with some energy, that great dark mass necessary to explain the linear velocity curve of spiral galaxies disappears, since the space-time itself, which has always surrounded the galaxies, would provoke the necessary gravitational effect.

What about accelerated expansion? For that we need to look at a little bit of what quantum chromodynamics says.

Quantum chromodynamics is the theory that attempts to explain the strong interaction, which is one of the four forces we know to exist: gravity, electromagnetism, weak force and strong force.

The strong force is what keeps matter in place and makes it possible for it not to disintegrate or rather vaporize. As we know, matter is composed of atoms. Atoms in turn are made up of a nucleus and its electrons. In a nutshell, the electron has a negative charge, and the nucleus, which in turn consists of neutrons and protons, has a positive charge. Since the protons have a positive electromagnetic charge and the neutrons zero, we can say that the nucleus as such has a positive charge (that of the protons). Here the main force is electromagnetic. We have the electrons spinning around the nuclei.

But if the electron is negative and the nucleus positive, why doesn't the electron fall to the nucleus since opposite poles attract? This is where we unfortunately have to resort to quantum mechanics. Electrons, and all particles in general, are only allowed to be in a few energy levels, that is why it is said to be quantized. It's as if I were an electron stuck in an elevator. I can only get off on the floors. Between floors in theory, I can't get off. If I go up to a higher floor, that costs energy, and in the world of the electron that energy is taken from a photon of light, for example. Likewise, if it goes down a level, it gives off energy by releasing a photon. In this simile, we could say that, if it continues releasing photons, it would finally reach the zero floor, the nucleus, but according to our hypothesis, when that electron reaches the first floor, it no longer has energy that it can release to go down further. It is in the minimum energetic state and that is why atoms do not disintegrate, which by the way is quite convenient for us.

But what happens in the nucleus? If neutrons are neutral and protons have a positive electromagnetic charge, two protons should not be able to coexist in the same atom, since equal poles repel each other. And so, it is, unless there is a force even stronger than the electromagnetic force that can hold more than one proton together in a nucleus. That force is the strong force, which is much stronger than the electromagnetic force, but whose range is very short and limited to those nuclear distances. In fact, the strong force that keeps protons and neutrons together in the nucleus is a residual force of what really happens inside the protons and neutrons, since these particles are not considered elementary because they are composed of other particles, the quarks. Electrons are elementary particles, and they are not subject to the strong force because they are another type of particle different from the quarks.

A proton is assumed to consist of three quarks. Likewise, each neutron also consists of three quarks. These quarks interact with each other in many different ways, but at these distances the most relevant interaction is the strong one.

I am here giving a few brushstrokes to all this history and this text does not pretend to be a scientific text, for that there are many other books that can explain all this in a way if not clear, sure that from a much more accurate perspective. From the concept of the atom to the quarks, there is a long history where many brilliant minds have contributed their granite, and where I have summarized it here in a grotesque and probably confusing way for those who are not familiar with all this world. However, all this has been to arrive at the idea hidden in quantum chromodynamics, which is the theory that explains the strong force at the quark level.

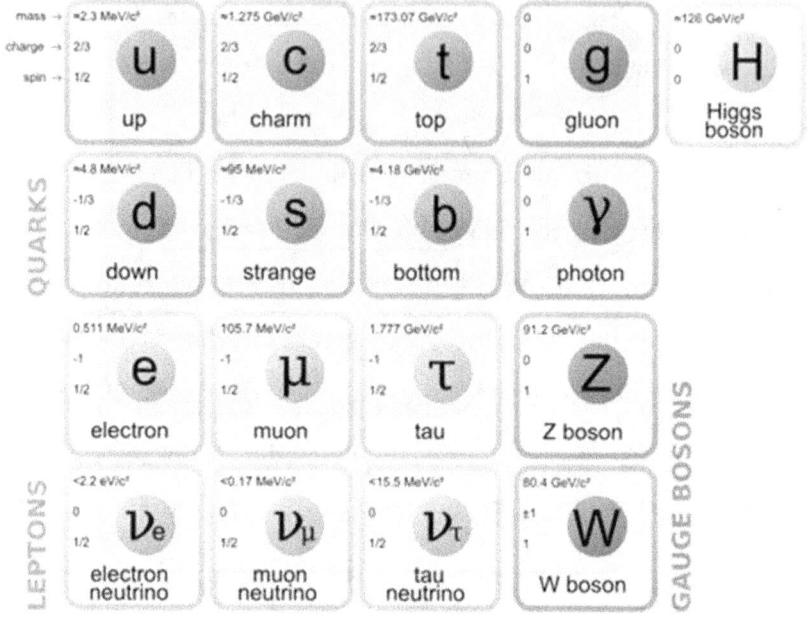

Illustration 4 - Standard model of elementary particles

We had been left with the assumption that protons and neutrons each consist of three quarks. As it is obvious, for the human being, we have also tried to dismember the protons and neutrons, and to observe the quarks naked and under the microscope, but we have not succeeded. We are not able to separate the quarks individually. Moreover, the more we try and the harder we try, the worse it gets. It is as if they were in motion, but stuck together, that's why the elementary particle for this interaction, the strong force, is called gluon, from glue.

The world of quantum chromodynamics is strange. These quarks have fractional charges like 1/3, we have to make use of a new property called color, and, finally, a series of curious inventions that make us think that this cannot be so. However, like quantum mechanics, quantum chromodynamics also works and is correct in its predictions, so, for the time being, let's leave it at that.

As I said, the more we try to separate a quark from its other two companions in a proton or a neutron, the more it resists to be separated from its siblings. Let's continue with the sibling simile. Let us imagine that we have three siblings. Like all children, if they are left alone and quiet, they will be playing their own games, destroying things and building new ones, and also, as a matter of course, hitting, insulting, biting, ... Suddenly, a strange adult enters the room, picks up one of the siblings and says that he is taking him away. In a real world, it is likely that he would succeed, not without certain difficulties put by the brothers, but in our simile, this would be impossible. The fun and also hatred that the brothers share when they are calm, transforms into love when someone tries to separate them. A magical love that transforms them into extraordinarily strong and muscular beings that makes it impossible for anyone to separate them.

Quantum chromodynamics tells us the following. Each proton or neutron is formed by three quarks. These quarks behave freely inside their small sphere called proton or neutron (or other particles, but we do not consider them here for the sake of simplicity). That is, the proton, as well as the neutron and other particles, is practically empty. It is like one of those plastic balls where you can put it in, inflate it with air, and throw it into the water. Imagine this transparent plastic sphere and inside 3 balls interacting with each other. Inside that plastic sphere they move with some freedom, but if they try to get out or someone tries to get them out, they bounce against the wall of the elastic plastic sphere and it makes it impossible for them to get out. The more you try to get them out, the more the elastic properties of that plastic sphere, which we should consider indestructible, will oppose.

Another simile that is usually given is that of the three spheres joined together by elastic bands. If these elastic bands are not very tense, the spheres, the quarks, will move with a certain freedom. However, if you catch a quark and try to pull it, the elastic cords will tighten and you will have more and more difficulty. In our world, those rubber bands would eventually break. In the quark world, those rubber bands are unbreakable, so the more you stretch, the harder it will cost you and you will never achieve your goal. So much energy is invested in this process that in the end all you get is that the energy is converted into matter and you have to start all over again. The energy is transformed into a pair of particles that combined their properties cancel each other out leaving only energy, but with that surplus of energy they are spontaneously created to combine with other particles existing in the surroundings. It is as if pulling a quark, in the end we manage to break the rubber, but that break that contains a lot of accumulated energy, in the breaking apt that energy is transformed into a pair of new quarks that are placed at the ends of the broken rubber. As a result, we still have a proton or neutron or whatever, intact, and we have created a gluon formed by two quarks by investing a lot of energy, but we have not managed to keep a separate quark.

In short, this strong interaction that tries to be explained by quantum chromodynamics, at tiny scales within what could be the sphere of a proton, behaves quite relaxed, and lets the quarks move freely. However, at "large" distances, and here large distances are those that exceed the diameter of a proton, this force acquires an enormous robustness that makes it impossible to break. As we have seen, it would only be to invest energy, to build more matter, but without separating those nuclei. It is like a father with his children in a park with swings, slides, etc. ... As long as the children are inside the park and in sight, there is no problem and the children move freely. But if one of them leaves the park following a butterfly, the hysteria starts for the parent who automatically changes color and goes after his child.

Another simile is sheep with a sheepdog. As long as the sheep are more or less together, nothing happens. If one or a couple of sheep seem to wander away from the group, they will immediately have the dog in front of them barking at them to come back to the group.

Therefore, this is a curious property of quantum chromodynamics: the "farther away", the stronger, and the closer, the more imperceptible.

Now and after this vision of quantum chromodynamics presented here without much rigor, let us return to the subject that had brought us to introduce this topic: the accelerated expansion of the Universe.

And what has one thing to do with the other? Well, nothing and everything. In the end, I am convinced that everything is related to the Whole within this adventure of ours that we call life, or simulation, or whatever it is that we experience.

Let us recapitulate. We had space-time to which we had endowed some energy. Another way of saying it is that space-time can be transformed into energy, and therefore also into matter. All interchangeable, space-time, matter and energy.

The famous single Amale element. We can even write some formulas. If we call space "x", time "t" and speed "c", we have:

$$c = \frac{x}{t};$$

$$x = c * t;$$

$$t = \frac{x}{c}$$

Suppose that the speed "c" is now the speed of light.

We also have this other formula:

$$E = m * c^2$$

If we substitute velocity for space-time, we would be left with this:

$$E = m * c^2 = m * \left(\frac{x}{t}\right)^2 = m * \frac{x^2}{t^2}$$

From which we can conclude that:

$$\boldsymbol{E * t^2 = m * x^2}$$

This formula surely lacks some parameter and it will be necessary to be careful with the dimensions, so I do not discard that the introduction of some constant is missing. It has simply been a quick exercise to combine the Amale in its different modes of presentation: space-time, matter and energy.

We have seen that matter and/or energy deforms space-time. If we endow space-time with intrinsic energy, now also space-time should succumb under its own existence, which does not happen, so we should infer from here that the equivalent energy of space-time should have some definite characteristics and possibly contrary or of opposite sign to gravity but whose scope and power is very small, so small that its energetic presence at galactic scales is more relevant as an energetic presence favoring gravitational fields and replacing the need for dark matter-energy, but that at very small scales, of the size of quarks or when we get into a black hole, that space-time energy with negative polarity, makes space-time not collapse by its own existence or in collaboration of matter and/or energy. Moreover, that negative energy of space-time, even being extremely weak, like gravity, does not have double polarity (as for example electromagnetism) and is accumulable. Therefore, having only one polarity and accumulating, and even being very weak, at cosmic scales it finally imposes itself and causes the accelerated expansion of the Universe that we see.

The above confirms that the previous equation is not correct or that at least it is missing something or some sign or something else.

We can also see space-time in the quark world where it can no longer collapse as a minimum energy state, so it can go beyond it, i.e., disappearance.

We would have here a kind of duality. On the one hand, we would say that space-time is "negative" and energy-matter is "positive". And their summation, null? Well, it could be, but that is already getting too much into the world of speculation.

15.   I am you.

When we open our eyes, this is what we see: matter, energy and space-time. And, in fact, although we believe that we are independent beings, if we think about it a little more, the molecules and atoms that in theory limit our being are in close contact with the molecules and atoms that surround us. The molecules of our eyeball or of our hand are in contact with the molecules of the air, and these in turn with the lamp we have half a meter away or with the tree we see through the window, with which we are also connected.

Everything is connected. Really, where does my being end? In turn, if my being is not limited, neither there is anything out there that is not me. What I think is my "I" becomes a tiny part of the whole. That "I" is in contact with everything, with you, with a cat, with a tree, with a rock, with a star, with everything. Everything with everything. That is why if we damage any part, we are ultimately damaging ourselves.

Curiously enough, this is what most religions promulgate: love your neighbor, do no harm. Other cults have gone a little further, such as Buddhism, and have managed to see this cosmic connection of the whole with the whole, and therefore suggest above all compassion in our judgments, because without knowing it, we are judging ourselves.

The world of dreams has also shown us that there are unexpected connections that also arise from some part of our brain that we do not know or understand. How is it possible that in our dreams we see things that we had not seen before or could not infer? Where do these ideas come from? In our dreams we mix up everything, living, dead, unborn beings, places, times, life forms, sizes, physical effects. While many, or perhaps most can be inferred from our experience in life, I fear that not everything is so clear. Sometimes, we have dreams where we see and feel that mystical connection of the whole with the whole, and it is something we are not concerned about. Rather, when we wake up, we are with a silly smile like some recently deflowered youngster. Exhilarated, content and somewhat more complete.

Obviously, many of the things I am writing in these lines are monumental nonsense, but what I want to get to is to be able to show that our world has some really serious metaphysical deficiencies, that we are taking many things for granted and this should not be like this.

We have stopped thinking for ourselves. We are like cows, with my respect for them. We are little children.

- *Daddy, daddy, why do I exist in this form and time?*
- *Eh... because the Gods have willed it.*
- *Ah... okay.*

A real child would go on with a long list of "and why", but we have been satisfied with the first explanation we have been given, and this should not be the case. We have too much work to be wasting our time. We have some serious problems to understand our world that must be investigated and why not, someday, solved. Our story is not finished, many chapters are missing, but it seems that we have run out of ink in our pens. We have to wake up. Quantum mechanics, dreams, virtual reality, black holes, simulation, Big Bang, expansion of the Universe, dark matter and dark energy, free will, ... Choose a topic and start thinking for yourself. Question the fashionable explanation. Stop wasting time. It is our moral obligation to investigate since we have no other choice. We can soar like eagles into the unknown, or crawl through the sewers watching what is broadcast on television. We must learn again to open our eyes and see what is in front of us. This could be one of our supreme missions and although many explanations are general and apply to all of us, many others are particular to each one of us. Not all general explanations serve to answer my particularities as the unique being that I am. And as the unique being that I am, no one will understand me better than I understand myself. And if I am not fully understood, how can I expect someone to come along and give me answers to questions that only I understand? It is I who must give answers to my questions.

I can have "small" questions and "big" questions. Small questions like what am I going to eat today? Or big questions like what is the meaning of life? For both types of questions, if the generic answer is satisfactory, then great, one less problem. But if the generic answer doesn't satisfy me completely, what am I waiting for? Do I settle for whatever they give me? Do I wait for someone to come and answer it for me?

These are the attitudes we take. Mysteries of the Universe, ok, let's put a deity there to calm our worries, even though deep down you know that deity is a sham, a hole plugger.

Well, okay, I'm not happy, but just as technology evolves throughout history, so do ideas and knowledge. That gap will be filled sooner or later. Yeah, fine, but when? Do we think we'll still be here when someone finally comes to explain? You wipe your own ass. You answer your questions. And if you have no answers, be brave, admit it and go to work. Look for an answer of your own, or at least a temporary explanation that satisfies you, knowing that you still have a long way to go and determined to fight and not leave that path until you have your answers.

It is the myth of our culture as humanity. The hero and the challenge. We are all heroes in search of our destiny, our answers.

So, we know the path, it is there, in front of us. Let's keep going. The task ahead may seem immense and difficult, right. But it is our obligation, it is our path, no one is going to walk it for us. Don't be lazy. Laziness is the enemy of fulfillment, of the feeling of being complete.

The task is long, but like everything and like all roads and journeys, it begins with a small step. The big tasks are undertaken in this way, the big task is divided into subtasks, and these in turn into smaller ones, so small that they are to our satisfaction and do not scare us. It is a long-distance race; it is a marathon. When you run a marathon, you know what you have ahead of you, 42.195 km. Look like that it seems crazy. Let's say a time of four hours to run the marathon. No one picks up and starts running like crazy to meet this goal. The distance is divided by the time available and this gives us that we have five minutes and forty seconds to run each kilometer. A comfortable pace we would say and the problem is already seen as something more manageable. And even if you get to kilometer 30 and you can no longer maintain that pace that you considered "easy" at the beginning, it doesn't matter because you know that you have already eaten three quarters of the problem.

You have to push yourself and finish that cake. Even if you ultimately don't finish, the sense of revenge is clearly born in you. If not now, it will be in the near future, but that pie will be mine. You see the goal. It is yours. What seemed impossible is within reach. And how did it all start? By dividing the task. And if you fail on the first try, you get up and try again. And if you fall down again, you get up again. As many times as it takes. No one said this is easy. And the one who falls, gets up stronger than the one who did not fall. Of course, you have to be attentive to the falls to see why you fell, learn from it, memorize it and not fall again for the same reason because that is a waste of time and experience.

In short, any problem and/or unknown is solvable if it is divided into manageable parts. If we fail to solve any of those parts, it does not matter, because that is also part of the solution, from which we will come out stronger.

Let's not get distracted, your problem is there, right in front of your eyes, solve it, and get to work now.

In a marathon you have a time limit to complete the distance, usually five or six hours. If your race pace is such that you will not be able to finish in that time, you are withdrawn from the competition. "The broom truck" picks you up to clear the track. In real life, "the broom truck" is death with its cruel scythe. In a marathon, you know what the time limit is. In real life, do you know what your time limit is? Negative answer, so minute that you are not taking advantage of, minute that you are definitely losing to solve your problems, because remember, for many of your problems there will be generic answers available and satisfactory, but for other problems of yours, the answers are yours to find. Everyone has their own way. Perhaps I can teach you the generalities of your path and how to walk it, but I cannot walk it for you. Neither I nor anyone but you can walk that path of yours.

It is like meditation. Today there are many forms of meditation, although the one that has taken the definition by a landslide is what we understand by meditation related to Buddhism. From this creed, even if we dispense with everything else, we should stay with this practice, since its benefits are proven. Even if we never reach the highest level and are with our mind's eye seeing everything clearly, the simple fact of practicing meditation is good for our character and our health. But to do meditation someone has to explain to us how. We sit (or lie down) properly, close (or not) our eyes, count our inhalations and expirations, go over all the parts of our body and relax, etc. .... And, above all, we try not to lose concentration by relaxing our brain so that it calms down. That means expelling that horde of crazy monkeys that do not stop bombarding us with new ideas, feelings, memories, ... This is how meditation begins, in a few lines I have summarized what others also try to explain in pages and pages. In this case, it is not a question of reading and reading to learn, but to know more or less the basic rules and to start playing. Moreover, this is not swimming, or playing soccer, where I can tell you if you are doing well or not, where I can correct you in your vices and defects to achieve optimal performance and results. Here, I can only tell you what most people do to meditate, what works for me or has ever worked for me, and some experience and/or feeling that I have felt and perceived, try to tell you about it, breaking with words the beauty of that feeling and experience.

Here, you alone have to meditate. No one can meditate for you. Likewise, no one can solve your particular problems for you. And likewise, it depends only on you if you practice meditation and include this practice in your routines, and it also depends on you if you decide to solve your problems, or on the contrary you lazily wait for someone non-existent to come and do it for you while a dark figure with a scythe approaches inexorably to wake you up from your lethargy and make you see how you have wasted your only chance (at least in this life and/or simulation...).

Moreover, all this (and as expected) is interrelated: problems, meditation... dreams... Even for those big generic problems of ours (accelerated expansion, black holes, quantum mechanics...). That is, we have problems, which seem to be unsolvable even by applying partitioning rules. Where can we get new ideas or clues that make us see the problem from another perspective and in turn allow us to solve it? Indeed, through meditation would be one way, and another alternative would be dreams. In both environments, we see the world with another perspective, which is perhaps even more valid and authentic than the one we believe in our first life as "conscious" beings.

Through meditation we try to see the world, or at least what we think the world is, without any preconceived concept, and without any kind of emotion. It is as if we were newborns, where everything is new to us, but without applying emotions to it. Whatever we see, we are just observing, without emotion and therefore also without judgment. Theoretically, therefore, through meditation and once we have reached a sufficient state on the ladder of meditation towards ecstasy, we should be able to analyze the problems for which we do not yet have a solution from a new perspective, which in turn leads us to the solution of those problems, or at least provides us with the next clue for us to continue working on those problems. The mention here of "clue" is important, because clues need to be seen, clues need to be interpreted, clues need to be recognized, clues need to be applied... And that is not always possible. For that you also have to train. We have to know how to train our own brain first to recognize and remember any hint of a clue, so that with a little more time we can work on that clue. This, and in the world of meditation and at least for me with my extra white belt as a mega novice, is not at all easy. How can I pretend to identify clues and remember them when at the same time I am trying to be like a rock that only observes? How can I run a marathon with its 42.195 km? The answer here is simpler, with training. And if we not only train, but also have a training plan, so much the better. This same answer is valid to answer the question of how to be and at the same time not to be during meditation.

Something quite similar can be said about dreams. As we have seen above, in the world of dreams anything is possible. We are not bound by the laws of Nature. We can fly, we can breathe at the bottom of the oceans and in the cosmic void, we can be another person or animal, anything goes for a while. The time I am dreaming is perhaps short, but as time also has its own rhythm within the dream, that time should be enough... As that capacity within the dream, we can effectively and not figuratively "think outside the box". That is to say, to get completely out of the environment of our problem and look at it as a whole, as if flying over that problem, perhaps acquiring a global perspective of the issue, or at least a new vision of the problem.

As I mentioned before, to solve a problem you have to break it down into manageable blocks, but not only that. In the real world, whatever that means, you also have to know how to distinguish between the wheat and the chaff. This is not about solving a mathematical problem in primary school where you are given a series of data and with those data and applying a series of operations, you arrive at the solution. No. Even in more advanced stages of Primary Education and later grades, we see that educators are introducing data in the problems that are useless. And we start to think about the malice of some teacher and we remember their mothers. But no, that's a golden lesson. To come across data that we know is worthless. To know how to distinguish them and leave them in the drawer of useless data, and it is better that this drawer does not go directly to the wastebasket, since, like everything else, maybe now for this particular issue it is not useful for me, but maybe it will be useful for some other problem that I have now or that the uncertain future will bring me. Knowing how to distinguish between the music of our problem and the background noise is a big step forward. And it is a big step forward because in this way we can focus our attention on the data that are important and useful for our objectives.

This is difficult in the "conscious" world, in the world of meditation, but perhaps even more complex in the world of dreams. First, we have to remember the dream, a subject I think I have already said something about, but I don't mind repeating myself. Basically, it is a matter of convincing ourselves that we are going to remember our dreams. It is, again, training our brain to perform a task. And with continued practice, it seems to work. What I have found is that you have to be consistent and persistent. That is, if one day you are so busy and/or entertained that you have forgotten to do the homework to remember your dreams, you lose previously acquired battlefield, for which you will have to stick again. If you don't train, as with running or any other sport, what you have gained is lost. And you have to notice that, in this case, the training is pretty basic. Things like saying to yourself "I'm dreaming", thinking before going to bed that you are going to dream and remembering... In short, tasks that are simple, but that, if we do not persevere on them, another day goes by without realizing that you have done nothing, and you have lost ground.

That was the first point, to remember the dream. The second would be to know how to interpret dreams and to extract from those interpretations the clues that could be useful to us for the resolution of our unknowns. But of course, here we enter again into a rather unknown field, the interpretation of dreams. That I dream I am sailing in a yacht for example can mean several things. That I want to be rich and have a yacht, to start with. But it could also mean that I need a vacation. Or that I feel lonely in the middle of the ocean, or, on the contrary, that I need some solitude for my things. Or that the new job or position I'm looking at as an adventure... And so, we could go on with a few more examples. To have a harder look, we should go deeper into that dream to find out if I was alone or accompanied. If I was accompanied, by whom exactly and my current situation with those people, if they are still alive or dead, or if they are unknown. If it is good weather with a beautiful radiant Sun and a calm sea. If a storm is approaching and the sea is rough. If I am moving or anchored. If I am near or far from land. What exactly I am doing. If it is day or night. What I am saying and/or thinking. Without details it is impossible to have a more or less accurate approximation, and to this must be added the unique and particular situation of the person dreaming. It may be that different people with different situations dreaming something quite similar, find a meaning of their dream disparate. In any case, as we see, this does not seem a simple subject and unless each of us study this subject of the interpretation of our dreams, we will have to go to another person, probably a psychologist so that once we have told him all our life with all our darkest secrets so that this professional has made a "basic" idea of the type of person you are, you can proceed with your dream, so that the psychologist interprets it according to his knowledge. He/she may or may not be right, because this professional is not you, he/she may be very good, but he/she is still not your person, with your history, your experiences, your emotions and your particular way of seeing the world. Besides, it is not a free service. Buy a couple of books on the subject and start doing this work yourself. The fallacy of ignorance that you have and that a psychologist may not have, will be more than covered by a better knowledge of the case, since we are talking about ourselves. Just to add a word of caution. Let us be honest with ourselves in these analyses. Let's call things by their name and without excuses. If you're fat, you're fat, don't fool yourself by saying it's temporary because of the holidays half a year ago. If you're bald, you're bald. You can comb your hair in different ways to disguise this, but the wind of truth will reveal your secret. If we are not capable of crudely self-analyzing ourselves, in that case, yes, I would recommend the assistance of a professional who can see us without such subjective eyes as ours can be, especially when they look inward.

In addition, the study of dreams and their interpretation should be studied or at least taken as a hobby for the mere fact of expanding our culture. But first let's review what has been said in the last paragraphs:

- To solve our problems, we can use meditation and dreams;

- In meditation we must be able to continue in the trance and at the same time divert our attention to our problems;
- In dreams we must be able to remember dreams and interpret them.

16.   Myths and Legends.

And now, let's comment a little about our culture as humanity. In all parts of the world there have been myths and legends. These myths and legends are all fantastic stories, where we can find beings with powers, gods, demigods, humanized animals, ... In all these stories there is usually a message of behavior, or a kind of lesson learned. These messages can be more or less clear. They can be hidden in the story, and they can also be summarized and captured at the end of the story in a kind of moral. Within these myths and legends, we can also find the full range of age classifications: we have stories for all audiences, children's stories, to genres only for adults with stomachs big enough to digest tales of parents eating their children, dismemberment of bodies for superfluous reasons, and so on. We have the availability of all genres that apply to cinematography. Thus, we have legends and myths that are dramas, comedies, adventures, even historical, romantic, science fiction, horror and terror, children's, pornographic, ...

All legends and myths tell us a story. A story as we have said with a message. And a story in a magical environment, full of fantasy. And I say: couldn't we apply this definition to dreams as well? Let's see. Dreams tell us a story, a story with a message and in a magical and fantasy-filled environment. Well, it's not the definition we're going to find in dictionaries, nor in Wikipedia, but it fits me pretty well too.

What I want to point up is that probably all myths and legends have evolved from some starting point, which I place in dreams themselves.

Someone has a dream and tells his partner or friends about it. If it is interesting enough, this dream passes from mouth to mouth and becomes known, even popular. As time goes by, and playing "broken telephone", this original dream, is transformed into a myth or a legend where embellishments and details have been added, even chapters, little by little, and practically without being its objective, unintentionally. With the participation of several people who have been adding more color and symbolism.

Today we may not see this so clearly, because in today's families there is not much dialogue. After spending all day beating each other with daily banalities, the day ends up relaxing in front of the television. When we are with friends, we don't get to that level of sharing our dreams in a serious way. We just laugh, drink some alcohol and talk about politics, sports, the opposite sex and our family adventures, we talk about the amazing world of living with our own children and how they seem like an endless box of surprises.

There is no more room for evenings where stories were told, awakening the imagination and preparing us for our own dreams. In telling those stories, we added new parts, changed parts, and were directors, scriptwriters, actors, sound engineers, special effects, lighting, sets, all at the same time. We played different characters with different voices and personalities. We are also no longer around a campfire day after day with a bunch of people, some known and some not, walking around every day, or warring and dying by the thousands at a moment's notice.

Perhaps before the sense of risk was much more intense, and not only risk, but the lack of security. This, together with the fact that we had to share stories together, must have opened up a vast field for sharing dreams and with them the word of mouth of the broken telephone and from there, to the great myths and legends.

Let us imagine that we are two centurions resting in an evening after six days of forced marches under the command of Scipio Africanus and that we are at the gates of Carthage. The next day we will fight to the death against the Carthaginians. Perhaps we will survive, perhaps an opponent's sword will pierce me, perhaps they will cut my head with the edge of another sword, perhaps an arrow falling from the sky and without a definite aim, will finally decide to pierce my right eye and reach my brain, perhaps a spear will fly swiftly to my chest, pierce me and leave me agonizing until I die nailed like a scarecrow.

Or maybe I'm a U.S. Army soldier deployed to England in World War II on a landing craft approaching the French coast on a day called D. Away from home, cold, seasick from the swell. Stuffed in that can of sardines along with a bunch of other sardines, almost all of them shivering, puking or pissing themselves.

These situations, thanks to all the Gods, are not every day, but for the people who lived them, they were real, starkly real. In those situations, a dream could be given a relevant meaning, endowing it with an important value. It is not the same to dream of dirty water today in Bilbao than to dream of that same dirty water on the South coast of England on June 5, 1944. In the first case, it can be interpreted as meaning that you are going to fall ill. In the second case, the interpretation would be "something" more funereal. But in those situations where there is, let's say, less security, the interpretation given to dreams is more valued and more vivid and real.

In the past, security was not the same, wars, human rights, diseases, ... Today, in most parts of the world, you go out on the street any day and there is a very high probability that you will return home safely. In the past, this was not the case. At any moment, your life could take a sharp turn or simply end. Protected by a fire at the entrance of a cave and surrounded by the rest of your tribe, you had spent one more night, but that didn't mean you were going to sleep one more night. Maybe you came out of the cave and there lurked a wolf, or a bear, or a rival tribe. Maybe you survived that first hour, and you had a chance to go hunting, but the easy prey in a last gasp, throws a hind leg that crashes into your head. Or maybe you are in ancient Rome or earlier in Greece with your security of certain position and wealth, and living in a city with a high wall, but a clap of thunder startles a pair of horses dragging a chariot and they run off in terror taking everything in their path, including your life. Or maybe you got out of those horses alive, but in a corner awaits you a poor wretch with a fierce hunger who gives priority to satisfying that hunger rather than his own life, a life already full of suffering despite his young age, and armed with a rusty knife he quickly ends your life to take a couple of coins and eat a hot meal with some wine before someone points a finger at him, and that meal mixed with the wine ends up in an open sewer because an executioner of justice has done justice by slitting the guts rather than the neck of this poor wretch.

Or maybe you are an Indian in South America fleeing from the culture and bloody rites of a neighboring people, fleeing at the same time from the dangers of nature, to end up fleeing from some new conquistadors eager to Christianize you or end your race for survival.

The degree of security that exists today in most of the world, no matter how much we complain about it, has nothing to do with what existed just a few hundred years ago, if not dozens of years ago.

If we put a few ingredients in our magic pot, we have the dish ready. Thus, if in this witches' and warlocks' casserole we add as the first ingredient a low degree of security, where survival could not be taken for granted, and where any clue and/or detail could mean life or death, and to this first ingredient we add dreams and their interpretations, suddenly we have a magic cocktail full of possibilities, possibilities that are given importance and relevance since we are risking our own life.

But the cocktail is not complete, we are missing a pinch of shared stories. Our own fears and dreams intermingled to have a story to tell in family or to our partners in hardship.

In short, not so long ago, we had three ingredients that made possible a rich field for myths and legends. These ingredients are:

- Poor degree of safety;
- Dreams and their interpretation; and,
- Dissemination of stories based on the above.

However, something has changed, it seems that all myths and legends belong to our antiquity with the exception of a few cases. So, what has changed? Well, let's take a look at the ingredients.

As we have already said, today the degree of security at a global level compared to a few tens of years ago is much higher, and nothing to do with that of a few hundred years ago.

On the other hand, we continue dreaming, there is no doubt about that, but first we do not make enough effort to remember our dreams, and secondly, when we remember a dream, we consider it anecdotal, maybe we tell someone about it, but it does not go any further. We remember a dream to practically forget it after a few hours without giving it any meaning.

Perhaps before remembering dreams was as complicated as it is now, but it seems to me that he who had a dream and remembered it, gave it a greater importance than it is given today, even if he did not know how to interpret it. Formerly, dreams were given a strong valuation and were not despised, and the destiny of people and even of nations has been marked by dreams. The when to attack of an army dictated by the favorable dream of a general, to throw oneself against the unknown into the thicket of a forest of trees or sea waves following a treasure or a meaning seen in a dream.

Today we can dream just as well, but before taking a step on the path envisioned in dreams, we prefer to keep circling in our cage of higher security recently acquired in our contemporary age.

Finally, the dissemination of dreams and stories. Before, around campfires or the trembling fire of a fireplace in a warm family home that in today's eyes would be nothing more than a hut. Storytelling, with all that it entails in terms of awakening our imagination and interpreting what we are told, has little to do with the current ritual of sitting in front of a television set, where they also project fabulous tales and stories, but which leave little if any room for speculation, interpretation and/or fantasy, since we get everything done, everything is already interpreted and told, and those television stories are fixed, frozen in time with characters with defined actors' faces. It is not the same that someone tells you the story of Alexander the Great even through something as fixed and immovable as a book than watching a movie about this historical character. It is not the same to read how he looked like and imagine it in your head, as it is to see Colin Farrell's face playing Alexander. It is not the same to read the description of the innumerable places where Alexander rode and that again you form an image of those places as you interpret them with that description, than to see them presented in a cinematographic scenography, no matter how good it is. The cinema gives only one vision, that of its director, the book gives as many visions as readers. Where is there more richness and breadth for interpretation? Where to see things in different ways, and thus tell them and that by the mutation of the "broken telephone" these stories are enriched even more?

Indeed, something has changed in the ingredients and, therefore, it is normal that today there are fewer myths and legends. And is this good or bad? Well, we will see, because now I want to introduce another topic derived from these myths and legends: religions.

17.   Religions.

What is the difference between myths and legends, on the one hand, and religions on the other? Well, many will say many, and may already put this simple question under something sinful. For me, however, and even if I am branded as sacrilegious, there is no primary difference.

In other words, of course there are differences, but these are of context and not of form. The differences are trivial, the background is the same, they are all fairy tales.

Let's start with the Greek and Roman gods. Not for any special reason, but because culturally they are the closest to my European upbringing. We have Zeus in Greece and Jupiter in Rome, with the same attributions: God of the Universe. We have Athena in Greece, and Minerva in Rome, being the respective Goddesses of Wisdom. And so, we can go on with almost a couple of dozens of gods. Poseidon and Neptune as Gods of the Sea, Dionysus and Bacchus as Gods of Wine, Aphrodite and Venus as Goddesses of Beauty, ... This is not the end of it, since we then have another list of so-called minor Gods and demigods, although these are less numerous. Here we have Eros and Cupid as Gods of Love, and Persephone and Proserpine as Goddesses of the Hells. But it does not end here, since it is normal to include another small list along with these

Gods and Demigods. This third list is that of the Heroes. We have Odysseus and Ulysses as Heroes of the Trojan War. Also, Heracles and Hercules, as demigod heroes. Each of these actors has its function and specific powers, and so while men warred among themselves, the gods also had their internal disputes in their Olympus, their mood swings, the favoring of one or the other on the basis of that fickle mood and the richness and exuberance of the sacrifices offered, ...

To anyone who does not have too many preconceived concepts, such as children, these adventures of the gods, demigods and heroes will seem like very nice stories. In that world of the Greeks and later of the Romans, mythology, legends and religions were in the same bag, and everyone saw them as something connected, although perhaps without explanation. We today have some plausible explanations that I have tried to convey in the previous pages. So far, I don't think there is too much conflict.

And now let us move on to Christianity, and I take this particular religion again for the same reason as above, i.e., my personal cultural proximity. What differentiates Christianity from the Greek and Roman religions? Several things undoubtedly, but from my particular point of view, nothing transcendental. Let's begin.

We went from having a multitude of gods to having only one all-powerful god. This makes certain qualities represented in certain gods suddenly disappear as they are no longer part of the one God. All the "good" characteristics of the old gods become characteristics of the one God. All the "bad" characteristics become ours because of our innate weakness. Thus, for example, love is good and comes directly from God. War is bad, but it is not the work of God, but of the vices of imperfect living beings, ourselves, who in turn are tempted by ministers of God who have fallen into disgrace, the demons, who are angels of God, who must also have had some imperfection to become demons... Angels and demons are not gods, but they could well be classified in the section of demigods.

And I wonder about the motive of that one all-powerful God to create imperfect beings, as well as imperfect demigods. Is it that this God does not pay attention to his work and does things anyway? Is he a careless being? Is the almighty God also careless? Is it a characteristic of divine perfection to be careless? Or has he made us imperfect to test us, or simply to amuse himself? If the latter, does not this morbid game, in turn, entail a certain degree of malice? But it is God, we had said that he was good... Well, someone explain it to me because if this God is not all goodness, but also has malice, we have just killed this fantasy called God and put it on the same level as any human being with his virtues and defects.

To the multi-gods sacrifices were made as offerings, normally we can be talking from wheat grains to multiple human sacrifices passing through an intermediate and more "normal" term of cattle sacrifices, such as oxen, sheep, etc.

And what is the sacrifice that Christianity demands now, for example? Well, self-sacrifice, not that we commit suicide, but that we self-flagellate ourselves for our "sins".

Another difference between myths, legends and also ancient multitheistic religions, is that these were fundamentally consistent, i.e., they had a beginning, a development, an end, lessons learned and guidelines on how to act. There could be fights between gods, too, but this is something that humans can see as normal.

Christianity, however, is full of contradictions, and it seems that this does not bother its followers. There is a common root for Christians, Protestants and Jews, and starting from the same, each one interprets as he pleases and they become enemies ready to kill and die for their particular interpretation, when at the root of all of them is the dictum of non-violence and love of neighbor above all else.

However, we have the crusades where in the name of God they killed without any remorse... Jesus, the son of God, freed animals, loved animals, opposed the sacrifice of animals and it can even be said that he was the promoter of vegetarians. However, today, in all the countries where these creeds are followed, huge quantities of meat are eaten. Meat from animals that are subjected to live a short life locked up in tiny cubicles where they can neither turn around nor lie down, and only wait to grow long enough for the time to kill and butcher them to feed the hungry jaws of all the followers of the prostituted words of Jesus and his God. And by the way, with today's livestock farming, we destroy the planet at an accelerated rate, and here we know with certainty the cause, unlike the accelerated expansion of the Universe, but that's another story...

These creeds also speak in their "sacred" stories about not worshipping images. However, the whole world is full of churches and worship centers in clear confrontation with the non-worship of images. The Catholic Church has a network of churches spread throughout the world for this purpose, while it stores unparalleled wealth in gold, diamonds and other goods. Economic wealth and political influence that dictated and dictates the future of countries with their millions of people and families. Wealth that is not distributed to the countries that need it and is displayed for their adoration. Wealth even where you have to pay an entrance fee to see it. I think that in the dictionaries with the definition of the word hypocrisy should be reflected some of this.

But well, let's leave aside these issues that can be quite conflicting, and let's focus on what has brought us here: the relationship between religions, myths and legends.

As I see it, religions, all of them, with more or less gods, are monothematic. Their only theme and sense of existence is due to death. Faced with the great mystery of death and in the absence of an accurate explanation, in order to calm our mind, and perhaps our spirit if it exists, we have no other choice, or at least so far no one has come up with any other way so effective, than to invent something. That something includes all kinds of Gods, as well as the homes they inhabit. It seems that life is more bearable if, faced with the inevitability of its end, death, we have a more or less fabulous and elaborate explanation that gives it a meaning. No matter how close we see death, we continue to deceive ourselves in order not to see what we have in front of our eyes. We can kill animals or simply watch them die. What do we see? Nothing, an inert body, which, if it is culturally acceptable in the corner of the world where we live, we eat it.

We can kill other human beings, or even see their lives slip away due to old age, illness, accidents or other reasons. What happens? Nothing. In certain cultures, in the past, this meat was also taken advantage of, especially if it was that of a powerful enemy and it was thought that their powers were going to be acquired through the meat. Catholic priests symbolize the ingestion of Christ by eating wafers and drinking wine, and they say "body of Christ, blood of Christ". Where does this symbolism come from? By eating your Christ, do you become as good as him?

What happens when a loved one dies? Again, nothing. We become sad because it has ceased to exist. The heartbreaking of this theft is somewhat softened the more self-deluded you are with some Kingdom of Heaven, since then your loved one is in another life, and probably, a better life. Incidentally, you tie this in with your own life and death, and become somewhat more at peace with a something beyond after extinction.

This is a self-defense of our brain against a possible collapse due to a lack of explanation of what our eyes see but we cannot accept and/or understand this futile existence.

It is like the number of conspiracy theories that have arisen with the coronavirus. Our minds would rather have any kind of explanation, no matter how absurd it may seem, than no explanation at all. How is it possible that, in the year 2020, as advanced as humanity is, a new virus could appear, infect millions and take away a million or so of those lives? It is not possible that these deaths, this theft of lives is the cause of a stupid virus that is not even visible to the naked eye... There must be another explanation. Some hidden interest of a country, or some secret organization with a clear purpose to kill old people, or keep us at home to achieve their goal. Surely, if we investigate and join forces, we will succeed in uncovering the conspiracy. I just hope I can do it before I fall off the flat Earth at one end... Obviously, I'm being sarcastic.

In the face of this, legends arise that explain everything, not only death, but also what awaits us afterwards. And here comes also something important and curious. According to most religions, if not all, what awaits you after death depends on your behavior in the present life. If you are good, to heaven, if you are bad, to hell. If you have behaved well, you will have a new life, if you have been bad, that new life will be that of a rat, if your religion speaks of reincarnation. With this we see that religions, therefore, dictate love and compassion in the present life on pain of punishment in the life beyond death. That love and compassion we also saw earlier when we were talking about dreams and meditation. In dreams and also in deep phases of meditation, we can sometimes see that ultimate connection between all things and living beings. They may be tiny glimpses, but they are there. They can also be fantasies and hallucinations, but in any case, they are there. For a religious person it is like seeing the face of their God. The great secret of the Universe and of our life is revealed. Everything is part of the whole and everything is connected. That connection means that if you hurt your neighbor, you are really hurting yourself, but in a deeper sense than the inherent psychological harm of remorse, for example. It would be something more like hurting yourself by cutting off a hand, for example. In that sense.

The ultimate connection of the whole with the whole that we perceive in dreams and in deep phases of meditation therefore suggests love and compassion during our waking life. To go against it is self-harm. This is not easy to see, or even to explain, and this may be the origin of the great universal success of religions. Besides giving an explanation to the death, and console with a future existence, the religions continue promulgating the love and compassion, not because this one is self-harming, but under threat of punishment in the hereafter for to see angry to your gods. If you believe in a god or gods, and they are humanized despite being gods, it is easily understandable to see a god angry if you have not complied with his or her dictates. And all gods are humanized to my knowledge. They all have attributes of humans, sometimes mixed with animals, and sometimes with more arms, but an arm is a human thing....

Religions are therefore legends. Legends as we have seen have their origin in dreams. In dreams we sometimes perceive the connection of the whole with the whole that dictates love and compassion. Love and compassion that in turn are also dictated in religions with a more accessible explanation for the populace, the punishment if it is not fulfilled. Fear is a very effective weapon that stuns and paralyzes us. Religions are elaborate legends that satisfy existential deficiencies by transmitting something that we perceive in dreams, even if we do not remember them. It is not surprising that religions are so successful and have moved the history of mankind.

And, if not this way, where do we go? That is, if we annul all religions or simply reduce them to beautiful stories, where are we? Well, if we strip religions of their power and move them to the corresponding level of legends, we have love and compassion as well, since both the legends and the vision of the connected whole have their origin in the same place, our dreams, our meditations, finally our mind. But of course, this vision without the support of a religion is not so clear.

If you do not believe in any religion, and if you are unable to see that universal connection, you are probably also unable to distinguish between right and wrong, between what is inherently just and unjust. In this situation you can take advantage of the rest of the world because you have no regrets like the rest, no regrets at all. Or so you think. Quick and easy profit will trap you in its self-feeding vicious circle, numbing you with earthly pleasures and even seeming to make you right in your actions. But sooner or later, your day will come, just as it comes to everyone else, in which you will wake up. Perhaps you will wake up on your deathbed, without too many opportunities to do anything anymore, but you will wake up. In the life we have, quick profit can bring us fame, riches and/or power. These are qualities we all crave and desire. We want to have an easy and carefree life. We want to dominate; we want to be heard. The only problem is that all of that doesn't calm our "spirit", whatever that is. We can be a king surrounded by treasures and his slaves. Or we can be a broker on Wall Street with millions in our account, getting drunk while gigolos dance for me. In this sense we can reach the top of the world in terms of fame, wealth and power. And what is the view from up there? Well, basically the same as from down here. Don't you wipe your ass alone? Aren't you alone in your head? Aren't you going down the tunnel to death alone? Metaphysical restlessness is not solved by fame, money or power. Only calmness, and with this we return to the beginning, being aware of the Universal Connection, or if you do not see it, calm yourself with one of the existing legends-religions that in the end promulgate the same thing: love and compassion.

To conclude this chapter, let us comment, even if only in passing, on the fact that nowadays myths and legends do not emerge at the rate they did in the past. We can also say the same about religions: no new religions emerge... Neither do we see the enlightened ones, divine messengers, ... now that we could videotape them so that their messages would be clear, with no room for interpretation and no excuses for us to kill each other in the name of a God. This is sarcasm, of course...

Today there are more books than ever. In them we can find an infinite number of adventures, which nevertheless do not become myths and/or legends. This overwhelming plurality has killed the possible singularities that could stand out to become myths.

On the other hand, if a book is worthy, it is almost no longer told by word of mouth awakening our imaginations, but we will read it and be satisfied privately, but for a story to become a myth or legend, it in turn has to be shared by a certain number of people, by word of mouth, through theatrical performances, ... However, nowadays, if something points towards that popular success, it becomes a movie which in turn kills any kind of individual opening towards imagination, as well as any possibility of it becoming a myth or legend, since that story/adventure has been reduced to a business, where if the result is good, we have a good time, but it does not transcend in our lives more than the approximate two hours of projection.

In this sense we could say as always that the present compared to the past is neither better nor worse, just different. Before, there were few stories and the best ones became myths and legends. Today, thanks to books, we have many stories, and this abundance makes it impossible to transform them into myths.

The proximity in time, and a greater general culture, I understand that they also have something to say. Nowadays, almost everyone can read and write. This means that we are not so gullible with everything we are told, because somehow, we have the possibility of corroborating that information by our own means.

On the other hand, if in our lifetime we have never seen a divine being, just like our parents and grandparents, the time continuum confirms to us that all these stories are pure invention or we have to put them in an unreal time for our understanding. It is very difficult for us to move back in time and imagine what our life would have been like 500 years ago. If we double time and put ourselves in 1,000 years? 2,000 years? In short, we can imagine visions, but it is very difficult for us to imagine how they prepared meals, how they cleaned themselves... Although there is a connection without temporal rupture, for our mentality it is as if we were talking about other worlds, and in those other worlds it is easier to imagine the possibility of including the "realities" of myths and legends with something more natural. Thus, we imagine the Minotaur in the labyrinth as a myth, but we do not see Godzilla anyway. No, Godzilla is within the genre of science fiction and does not pass into myth. Maybe in 2,000 years it will be, who knows?

The same would apply to religions. New religions do not emerge. The last one I am aware of is Scientology, which tells us that we are immortal spiritual beings, that we are basically good, that it is against wars, ... In short, a correct and healthy philosophy of life, but which also falls within what the rest of religions. The immortal spiritual being seems to pass from body to body when it is exhausted and dies. Okay, we have also seen that in other cults. Now comes the best part. According to Scientology, Xenu was a dictator of the Galactic Confederation, who 75 million years ago brought billions of people to Earth in spaceships. He then landed around volcanoes and annihilated them with hydrogen bombs. Their souls gathered in groups and attached themselves to the bodies of the living, and it seems that our souls and spirits are not really ours, but those of these extraterrestrial beings. That is, we really are them. This part already sounds to me like someone smoked something. To me, and almost everyone else, this part sounds like a fantasy that could not be seriously considered. However, I give the same vote of confidence to when God wrote the ten commandments on two stone tablets on Mount Sinai and gave them to Moses... Before these stories, I am dumbfounded, only to laugh at the thought that anyone in their right mind takes these stories as real.

However, the story of Xenu is believed by almost no one, and that of Moses by thousands, if not millions... Why do so many people consider the story of Moses to be within the truth, and so few believe in the dictator Xenu? The same explanations as above can be applied, i.e., greater general culture, and time perspective. With our cultural background we are adults and do not believe everything we are told. On the other hand, the story of Xenu arises in the decade of 1950, while that of Moses we should go to the thirteenth century BC, ie, Xenu is in the narrative just a few tens of years, while for Moses we have to talk not tens, but thousands of years.

As has been said, there is a temporal continuum, but the epochs so distant in time imply a temporal rupture, at least on a psychological level.

Besides, this scientology being true, it doesn't explain anything... Where did Xenu come from? Dictator of the Galactic Confederation? What Confederation? It's like if we create a religion based on Dark Vader. If we see all the Star Wars movies, we will know where Dark Vader came from, but the young Anakin

Skywalker will remain a mystery, since he looks like a more or less normal young human, who even having the power of the force will have his important metaphysical problems, unless that force directs us, as we can intuit, even in the Star Wars movies, towards a Universal Connection of the whole with the whole. The problem here is that we have all seen Dark Vader in the movies and his story is well told. This takes away many points for this beautiful story to become a myth or legend, let alone a religion.

Of the poor dictator Xenu, nobody has made a movie yet... Of Moses, yes... and the truth is that, if you see a movie of Moses, it takes away all the fun that the story could have. I'm not going to say it's funny so as not to offend, but it's a pretty grotesque representation for a supposedly divine event...

In short, there are reasons, as it could not be otherwise, for us to have a differential between myths, legends and religions, between contemporary times and past times. Some of these reasons I think I have exposed, and I am sure there are many more.

Throughout this part, we have seen again that, if we stop fooling ourselves, are brave and surrender to the obvious, we really have very few explanations. Necessary explanations for important questions. Important questions that we deliberately ignore by wrapping ourselves in the shelter of daily trivialities in order to hide our ignorance. But that ignorance does not make the important questions disappear and perhaps only when cruel death comes to cut short our lives, we awaken fleetingly to fall into the arms of despair.

But there are alternatives. To be humble and acknowledge our ignorance, if not to everyone else, we should be brutally honest with ourselves. Or do we also play a false role for ourselves?

From the bottom of the well of this ignorance of ours we must look up and see how some ray of wisdom sneaks in.

I see a ray that is indicating to me that all teachings derived from myths, legends and religions are the fruit of human minds. I see another ray that shows me that these human inventions have much to do with dreams and deep meditations where thought deviates from the marked path and enters through passages that end in doors to new worlds. I also see another ray that shows me a constant that is difficult to see, but present in everything we see and feel. This constant is reflected in myths, legends and above all in religions. This constant is perceived in dreams and meditations. This constant is the Universal Connection of all with all that dictates love and compassion.

That is the starting point, and it is a big step, but we have to enter that world and study it to finally satisfy our yearning, and each of us must walk that path alone, otherwise, all this that I am telling you stops giving explanations to the big questions and they are just empty words.

Finally, if we finally open our eyes and see what is in front of us, we will begin to see. Just as in dreams and also when meditating we fantasize; we also do this while awake. Our senses deceive us, and that distorted vision, so we transform it further through our emotions, memories and cultural stigmas, to end up, seeing something completely distorted. In this sense, and without going into a clear version of the simulation theory where we are only the result of some monumental computer program, we could say that we are talking about a light version of the simulation theory, in which our senses and emotions are in charge of presenting us with an a la carte reality, based, of course, on a "real" external world that we are unable to see in its fullness.

We have also commented before on how to try to stop self-deception. Basically, it is to take an attitude that we will call rock attitude. I am an impassive rock that observes without emotion. Meditation again is one of the ways, if not the way to get or try to get that rock attitude and see the world as it is.

To discover if the light simulation, which I am going to presuppose with all that has already been said is real, is in fact the heavy simulation, and this will be the one in which someone takes us for his toy and/or experiment, is not at first sight something simple. From here we would return to describe the "failures" of our physics and the compression of the Universe, with its explanatory gaps: quantum mechanics, black holes, dark energy and dark matter, ... But before reconfirming our ignorance and giving explanation to all our problems by saying that this is because it is a simulation, but with some programmer's conceptual errors, let us give ourselves some more time, because otherwise, a simulator in the end can take the role of an all-powerful God at least for our Universe, and even then, we still have no answer to know where that simulator or that God/Gods comes from.

We need more time, although it seems to me that with the problems before us, we have reached some important turning point in the understanding of our Universe, a point that undoubtedly requires a major Revolution, cognitive revolution, scientific revolution, or a mixture of both, and as I said, perhaps the first doors are in our own brain.

18. E.T.

Let us now turn to a topic we have not touched on: interstellar travel and extraterrestrials.

Let's start with extraterrestrials. As Feynman wondered, we can also ask ourselves: Where are they? Why don't they visit us?

Well, let's go step by step. Between an extraterrestrial and a homo sapiens, although minimal, there must be some similarities. Let's try to locate them.

First, they should be some form of life. I think that is self-evident. We don't care about their political system and their reproductive system, if they have anything like that, but they should be living things. If there is no life, there is no purpose. A rock can go its entire existence doing nothing, other than wearing itself out in the rain, and even if it has an accurate view of "reality", it does nothing.

With life, needs and purposes arise. Food, drink, grow, reproduce, this is what seems to define life. A cell wants and needs to feed, grow and reproduce in order to evolve, even if it does not know where to go. We also need to feed ourselves until we reach a stable and self-sufficient structure, surpassing babies and children, to reproduce and start the cycle.

If life becomes even more stable there is idle time for entertainment. With this, cultures emerge, and perhaps in some cultures, technologies will emerge that can take us to the stars. But let's take it one step at a time, because we're on a roll.

As we said, the first condition is that it must be a way of life. Second, it must be able to learn by itself and that this learning in turn can be transmitted to its descendants. This obviously entails a certain degree of intelligence, but perhaps not the intelligence that we define as such.

Today, a computer program can learn by itself and store this new information or learning on a hard disk, but for the moment, that computer program does not reproduce itself... But I suppose it will come. At that time, will a computer program capable of reproducing itself, learning and therefore evolving in some way, become a life form? Anyway, this question requires a chapter, if not a new book.

But it is not only learning that is necessary. It must remain and accumulate throughout the generations. It would have been of little use if our ancestors had discovered how to make fire, and had not passed it on, generation after generation without revealing the how. Knowing how to make fire is not written in our genes. Neither is it written in the genes of some monkeys to use a straw to extract ants, nor is it written in the genes of some birds to pick up stones with their beaks and throw them against the bark of juicy fruits. Such learning can be transmitted from generation to generation, but if there is nothing else, sooner or later, many discoveries will be lost.

And that is where another of the human geniuses, writing, comes in. Without writing and all that it entails, the accumulation of knowledge would have been laughably scarce. Writing is our way of accumulating learning.

So, extraterrestrial beings should have something like writing. Why might someone be wondering? Well, the answer is perhaps not clear at this stage, but writing that accumulates learning is necessary at later stages as we will see below.

Perhaps the most basic form of writing, and one that has emerged independently in a few cases, but more than one, is mathematics. The need to count seems clear. It is not the same to feed one child as two, it is not the same to feed a tribe of 10 as 20. It is not the same to pick 2 plums as 50, 2 baskets as 50. Counting things, units of something has been around for a long time, and when humans began to settle, and trade began, this counting exploded, and the need to keep some kind of record, too. And we can write numbers somehow, but those numbers have to refer to something, so we also need to write words. We already have writing. From economic transactions, we move on to writing more things, thoughts, philosophical treatises, ... Maybe one thing doesn't lead to the other directly, but having a multitude of writings with accounts, experiences, thoughts and learnings, I think leads sooner or later to some kind of technological revolution. And without a technological revolution, it is very difficult to build spaceships out of nothing. Hence, writing accumulating knowledge seems indispensable for our potential aliens.

These extraterrestrials of ours must have had enough time therefore to develop and evolve from simple life forms to complex life forms, to the point of developing the technology to allow them to travel through space. That in turn requires a habitable planet revolving around some stable star for billions of years.

On this subject there are thousands of writings with more scientific foundations than I would like to undertake here. We could, for example, take Drake's formula for estimating civilizations in our galaxy. The equation is this:

$$N = R^* * f_p * n_e * f_l * f_i * f_c * L$$

Where:

R* - would be the number of stars formed each year in the galaxy;

f_p - would be the percentage of stars that have planets;

n_e - would be the average number of habitable planets per star;

f_l - would be the fraction of such planets that would actually develop life;

f_i - would be the fraction of planets with life where they evolve into intelligent species;

f_c - would be the faction of such intelligent species that develop technology capable of emitting radio signals; and,

L – would be the average time an intelligent civilization capable of emitting radio signals could remain active.

This formula, as we perceive it, is to detect radio signals from a possible extraterrestrial civilization. Without going as far as the possibility of interstellar travel and visits to our planet.

Depending on the value given to each of these variables, we get one value or another, but we always get a positive value, unless we put zero to any of these variables. That means that we have a positive value, and we should be able to detect some signal, which unfortunately we have not detected.

Before moving on, let's discuss why this study is limited to our galaxy. The Andromeda galaxy is the "non-minor" galaxy closest to Earth. In fact, it can be "seen" with the naked eye. Together with 30 other galaxies, including our own, it forms the Local Group. It is speculated that before the total disintegration of the Universe by its accelerated expansion, our Milky Way, as we have named our galaxy, will collide with Andromeda. But that will be in about 5.86 billion years. In any case, right now this Andromeda galaxy is

2.5 million light years away. Since possible radio signals travel at the speed of light, a signal from Andromeda would take 2.5 million years to reach us. This is a long time. If it would reach us today, and after deciphering and analyzing, we would answer, our extraterrestrials from Andromeda would receive our answer after another 2.5 million years, if this civilization still exists, given that since they emitted, 5 million years would have passed, during which, many things can happen ...

For this reason, perhaps, in the first instance, it is better to concentrate on something closer to home, i.e., our own galaxy. Undoubtedly if we receive signals from other galaxies, it would also be a huge revelation, to know what intelligent life has developed in another galaxy, but unless, with the first message, the whole story would be told of that civilization, their concerns, desires, technological milestones, etc. .... will be difficult to get much benefit.

We humans, the first radio message capable of leaving the Earth with sufficient power was Hitler's speech at the 1,936 Olympic Games... Let everyone make their own guesses about the image that some possible extraterrestrials could form of the human race. Maybe nobody wants to interact with the giant virus that is killing their own planet...

As we have said, let's concentrate on the nearest, our Milky Way, which has a diameter varying between 150,000 and 230,000 light-years. Let's take 200,000 light years on average. One light year equals 9.46×10^{12} kilometers. This written plainly is 9,460,730,472,580.8 kilometers. If we multiply this by 200,000 light-years, the diameter of our galaxy, we get 1.89×10^{18} kilometers, that is, approximately 1,892,146,090,000,000,000,000 kilometers, which would be, therefore, the average distance between one end of our galaxy and its diametrically opposite.

These are still huge numbers. Let us be somewhat more optimistic and consider 1/4 of this figure, assuming that there are several civilizations in the galaxy and that they need not all be diametrically opposite in the Galaxy to our location.

In this case, we would be talking about 50,000 light-years. And that is still a lot of years. We are located on the outskirts of the Milky Way in one of its arms. Let's make a small sketch.

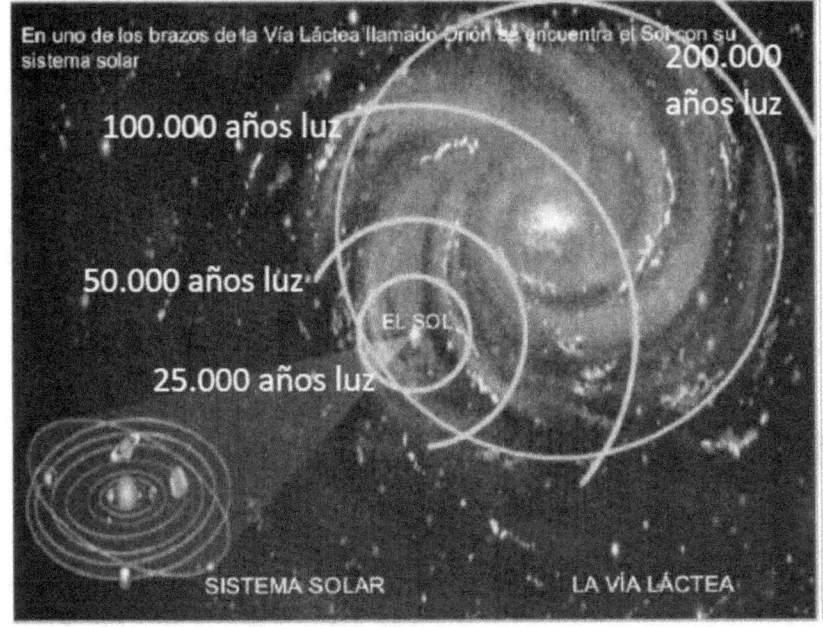

En uno de los brazos de la Vía Láctea llamado Orión se encuentra el Sol con su sistema solar

200.000 años luz

100.000 años luz

50.000 años luz

25.000 años luz

EL SOL

SISTEMA SOLAR

LA VÍA LÁCTEA

Illustration 5 - Distances

If we limit ourselves to 1/8 of our Galaxy, i.e., 12.5%, we would be talking about 25,000 light-years, therefore 25,000 years for a radio signal. By reducing the area, obviously the chances of finding civilizations capable of emitting radio signals decrease considerably.

Let's put 25,000 years into historical perspective. 25,000 years ago, a village consisting of huts built of rock and mammoth bones is founded in an area of the present-day Czech Republic. This is the oldest permanent human settlement archaeologists have found to date. It would not be until 18,000 years later, i.e., 7,000 years ago, that the wheel was invented. Some 1,800 years later we would have the invention of writing, i.e., 5,200 years ago. The first radio telescope was the 9-meter antenna built by Grote Reber in 1,937 that was built in the courtyard of his house, which coincides approximately in time with the first radio emission powerful enough to go out into space. That is, we are able to transmit and receive, considering that I am writing this in 2020, since approximately 83 years.

The human being since then is transmitting chaotically and without predetermined objective in a continuous way. As for our search aimed at this goal, i.e., to look for signs of possible extraterrestrial civilizations, we have to go back to the beginning of the SETI program (Search for Extra Terrestrial Intelligence) whose first projects emerged under the sponsorship of NASA during the 1970s, i.e., 50 years ago.

Now let's compare 50 years, versus 25,000 years for 12.5% of our Milky Way. Taking these data into account, the probabilities of receiving any radio signal are getting smaller and smaller, although not null. It is not surprising that we have not received any message. And the day we do receive one, because the odds, though small, are there, the outlook will not improve much. Why won't it improve?

Well, obviously it will be a historical event that will confirm or disprove many things and conceptions, and we will probably never be the same. But apart from the news, anything else?

I have to suppose that the first transmissions that we manage to detect will not be transmissions with the clear objective of contacting us, but some fortuitous transmission when that extraterrestrial civilization acquires that potential. Just as humans did with the Olympic Games in 1,936. Perhaps after a few tens or hundreds of years, they will be already directed messages, where that civilization will discover its secrets, its biology, its technology, its history, its world, its achievements, its fears, ... And we will continue receiving transmissions for centuries and centuries.

But we will also find out where they are. Let's imagine they are only 50,000 light-years away... It is possible that when we are jumping for joy to receive their message, even if it is fortuitous, this civilization on their planet has been extinct for 7,000 years, when we invented the wheel...

It is like 500 years ago when communication between the New Continent and the Old was done by letter that, loaded on ships, had to cross the Atlantic, but expanded to a scale that does not fit in our heads. When the news arrives, it is already too late. If the civilization still exists, but is in danger, because of pollution or because its star has become unstable, we will not be able to do anything, since when we receive that call for help, everything will have already happened a long time ago.

This also applies to us if we are the ones to launch a distress call....

Even in an ideal situation of stability, normal communication as we understand it is not feasible. If the civilization is 50,000 light-years away and we receive a message, we could send them our reply saying that we are here. That civilization would receive the message after 50,000 years, and if it is still there, and deigns to reply, we would receive confirmation after 100,000 years. It's not a very fluid dialogue to say...

If we were to receive a message, instead of attempting a dialogue, we should take concrete actions based on the available information, which, on the other hand, as we have seen throughout these pages, is both abundant and scarce. Abundant because of what we have accumulated throughout our history, little because we know that we still have a long way to go. The actions and decisions to be taken are basically the following:

1. Send them a message telling them that we are here, along with an encyclopedia with everything we know, our evolution, our history, our dreams, and that we are glad to be "neighbors". I suppose some will say that we shouldn't show our cards, that it's a bad strategy, and in

most situations, I would agree, but not in this one. If that civilization poses a danger, based on our knowledge, they have no way of getting to us in a quick way, so we would have "time" to prepare for any "eventuality". But in order to maintain a consensus, a safety distance could be put in place to send all our secrets, or not. The disclosure of fundamental "secrets" should therefore be inversely proportional to the distance.

2. Depending on our own technological advances, send a probe, or better yet a colony ship with families of humans for a journey of no return.

The fruits of these two actions will not be tangible in thousands if not millions of years, but is there any other option?

As we have seen, receiving signals from another civilization is quite difficult because of the brutal distances involved. But that's not all, since as we have seen, these distances imply tremendous times, and here comes into play another variable not contemplated so far. This would be the temporal coincidence of the existence of civilizations. Or at least temporal coincidence of emission-reception of the existence of civilizations. As we have seen, we have only started looking 50 years ago. It is of little use if a civilization was retransmitting non-stop and that those messages reached the Earth since the Triassic dinosaurs disappeared, 65 million years ago, until, for example, the discovery of the bacterium responsible for tuberculosis by Robert Koch in 1,882. Messages that do not reach their addressee, because the addressee is not yet ready. They would be lost civilizations.

Likewise, we have been broadcasting non-stop for 8 decades, we could have life forms here "close by", 20 light-years away, but maybe that life is still in the form of simple cells, which need a few million more years to evolve to discern our signals, but by then, maybe we are no longer here and we will become another lost civilization.

This last variable of temporal coincidence again reduces the probabilities of detecting signals from extraterrestrial civilizations. It is not that we are alone, because positive probability will always exist, but we are alone and separated from other possible civilizations by time and space, in quantities that far exceed what our brain can assimilate.

19.   UFO

So, if communication via radio signals is unlikely, what about space exploration? Well, basically the same probabilities, but reduced somewhat because we must take even much larger time scales.

The closest star to our Sun we have named Alpha Centauri. In fact, it is a triple star system, Alpha Centauri A, B and C. B and C even have a planet or two. This star system, which is the closest to ours, is only 4.37 light-years away, or 41.3 billion kilometers.

A Boing 777, has a cruising speed between 800 and 900 km/h. I put a commercial aircraft to have a reference known by most people who have already traveled by plane. So, let's assume 1,000 km/h to round up. Such a plane would take 4.13×10^{10} hours, i.e., 1.72×10^9 days, or 4,719,565.32 years. 4.7 million years for an airplane.

Obviously, a possible spacecraft would travel at a much higher speed. Voyager 1 was launched in 1,977 and is currently the farthest spacecraft from Earth at a distance of 21,093,299,768 km from the Sun, but it still has approximately 17,702 years to leave the Oort cloud, and thus finally leave the outer limits of the Solar System.

Photo 1 - NASA - On February 14, 1,990, the Voyager 1 spacecraft looked out into the inner solar system to take the first pictures of the planets from its location, at that time beyond Neptune. The dot inside the circle points to Earth.

However, it is not the fastest ship, but "only" the fourth.

Rosetta traveled at about 108,000 km/h, Helios B reached 252,900 km/h and the Parker Solar Probe has already reached 324,000 km/h on its first approach to the Sun in 2008, and is expected to reach 700,000 km/h at its closest approach to the Sun in 2025.

Photo 2 - NASA - On July 19, 2013, taking advantage of the fact that the Sun was at Saturn's back, this photo was taken. The arrow indicates the location of a small dot, the Earth.

Let's recalculate with this new figure. At the Parker probe's maximum speed, it would take 59,061,988.9 hours to reach Alpha Centauri, or about 2,460,916.2 days, which gives us 6,816 years. Wow, this figure we understand. About 7,000 years, which is when we invented the wheel until today.

If we were to launch a probe towards Alpha Centauri, after a few tens of years accelerating through the Solar System with successive approaches to important gravitational centers, mainly the Sun, this probe would reach Alpha Centauri in about

7,000 years, assuming that it has not crossed paths with some destructive particle during its journey.

It would reach Alpha Centauri, and overtake it at its breakneck speed to continue on into the darkness. In that fleeting encounter and if the probe's systems are still functioning it would transmit data to Earth, which would take 4.37 years to arrive. It would be the 9024 of our epochs if we were to launch that probe in this decade or so.

It would be complicated if our dizzying society kept a record of this probe's journey and that in 9024 someone is monitoring this issue.

The option of braking the probe at Alpha Centauri when it arrives there falls by its own weight, since this option would require providing the probe with engines and fuel, and this makes the mission much more expensive. Those engines and fuel would be used not for braking, but for guiding the probe towards gravitational centers with the object this time of braking instead of accelerating. But those guidance instructions would have to be given by someone, and that someone on the Earth of 9024 may not exist. The probe could be provided with the artificial "intelligence" capable of making the necessary calculations on its own to activate these small self-propulsions to finally manage to stay in the Alpha Centauri system with a radius to be defined.

And what would we get? See other stars up close, maybe other planets of those other stars. Make checks and reaffirm our science, assuming we have anyone on Earth who will still listen.

However, on the Earth of 2020 we are still so presumptuous that we assume that by 9000 we will be capable of much more and that we will not have to wait so long, considering the pace at which our technology is evolving, not to mention other real and closer problems, such as inequalities and the destruction of the Planet...

And so it is, in these millennia, we will be able to achieve many things if we do not destroy ourselves, but the distances we are talking about are still what they are, and the times may be reduced, but they will still be enormous.

Let's talk about something closer to home and look at its problems: Mars.

20.   Mars.

Photo 3 - Mars

The distance between Earth and Mars varies considerably. The maximum distance is estimated at 402.3 million kilometers, while the minimum is reduced to just over 57 million kilometers, which is when both planets are aligned and on the same side with respect to the Sun.

It should also be considered when this situation occurs since the minimum distance is reached when the planetary orbits are at their closest point to the Sun.

Regardless of this, let's keep in mind that the probe launches of a possible manned mission will be carried out taking advantage of these "windows" of planetary approach, obviously taking into account the relative velocities of the planets and the speed that the spacecraft can reach in order to perform the minimum trajectory.

This means that spacecraft are launched before that maximum approach between planets, for example.

Since the 1960s, exploratory probes have been launched to Mars from Earth, and we can confirm that it takes about 7 to 8 months on average, with many variables in play.

It could be achieved before and after, in any case, we are talking about months and normally it should always be less than a year comfortably. In fact, under the best conditions of distance and gravity, it is estimated that a round trip would be between 400 and 450 days.

In any case, once there, the decision must be made whether it is a fleeting mission like the 450-day, round trip with no time for anything except nailing some flag, or whether it is something more serious given the human and economic cost involved.

I understand, that despite the inherent risks it should be a bolder mission capable of remaining on Mars until the next favorable   conditions of maximum approach between planets.

The Earth and the red planet align with the Sun in opposition approximately once every two Earth years. There, stay and return is about 3 and a bit year. Let's assume 3.5 years. Well, it's possible. Obviously, we will land there and it will be a historical milestone. We could even colonize such a planet over the centuries.

There are many difficulties in these first manned missions. For a 3.5-year journey, you have to carry food and water, even if most of the water is recycled from urine, as is currently done on the Space Station.

It is necessary to provide an atmosphere for those 3.5 years, air that the astronauts can breathe, with machines that can recycle as much air as possible.

Enough fuel to leave Earth's gravity, leave a module orbiting Mars, while a sub-module descends to the red planet, as was done with the Apollo missions landing on the Moon, and then return to Earth, take off from Mars, and leave the gravity of this planet.

Getting all of this from the Earth's surface to orbit is already a major achievement and a huge cost, but it is possible.

We also have psychological problems of being cooped up, and getting away from everything you know and your loved ones. Astronauts in addition to their physical capabilities, intelligence and training, will have to be extremely stable on a psychological level.

Here I open a small parenthesis to mention that if what we are going to export to the ends of the Universe are psychologically stable astronauts, perhaps this is not a good representation of what human beings really are, with their madness, psychosis, artistic outbursts, making the right decisions, but not always the most logical ones, etc. ....

The list of difficulties for space exploration is long and costly, although almost all points are solvable.

21.   Spaceship.

However, there are at least two serious problems that at the present stage prevent space exploration from being as comfortable as the science fiction movies portray it to be.

The first serious problem would be the lack of gravity. The human body is designed to work in conditions of terrestrial gravity. The acceleration of gravity on the Earth's surface is 9.807 m/s^2, while on the Moon it is only 1.62 m/s^2, and on Mars 3.711 m/s^2. That is, the Moon's gravity is 1/6 of the Earth's gravity. Likewise, Earth's gravity is about 2.65 times that of Mars. If we colonize Mars, the body of the Martian human being over the generations will adapt to this reduced gravity for us earthlings, and the result will be that we earthlings will be 2.65 times stronger than the Martians. In a fist fight, a Martian would have nothing to do against an Earthling.

In the vacuum of interstellar travel, we don't even have that mini gravity of the Moon, we have no appreciable gravity. Our body is not designed for these conditions. All of a sudden, all the blood rushes to our heads, our heads swell, our muscles atrophy no matter how much exercise we try to do in weightlessness, our bones lose density, and a whole list of metabolic and physiological changes.

Even so, Valeri Poliakov remained 437 days continuously locked in the MIR space station on a mission that began in 1994. He was followed by Scott Kelly with 340 continuous days on the ISS station in 2015. Those 340 days were enough for him to orbit the Earth 5,440 times. The space stations actually have some residual gravity or mini gravity, but it is negligible. These spacecrafts, like satellites, are in free fall around the Earth (see Illustration 1 - Entering orbit.). Thus, future astronauts to Mars will arrive on the planet quite weakened after 7 to 8 months in weightlessness. At least, the reduced gravity of Mars will be favorable to them. This will not be the case when they return to Earth, after that first trip, 2 years in reduced gravity and another 7-8 months in weightlessness. When they return to Earth, they will feel extremely heavy, and their atrophied muscles attached to their weakened bones will barely be able to move. They will have a grueling recovery process to become world heroes...

And now let's look at a hypothetical trip to Alpha Centauri. What would happen to the human body after 7,000 years in weightlessness? A biblical generation is 100 years, but an effective and biological one is about 25-30 years, so we have 3 to 4 generations every 100 years. 7,000 years divided by 30 years per generation is 233 generations, where the human body will adapt generation after generation to the new "normal" conditions of weightlessness. Maybe not after "only" 7,000 years, but given enough time, the skeletal structure will disappear since we do not need a skeleton in weightlessness. There will be some cartilage to attach the few muscles that are necessary for life in weightlessness. We will be more worms than humans. "Homo Vermiculus".

After that enormous journey, any gravity will be deadly, so these future explorers will have to wear a very special space suit. They will have to be exoskeletons as well as maintain those creatures that in our current eyes will be extremely weak and vulnerable.

The second serious problem is radiation. Through space travel a lot of invisible particles to our eyes, but that can destroy us inside and out. This radiation is produced by our Sun, and the rest of the stars we see and do not see, by stellar explosions, collisions of galaxies, black holes, ... and practically every energy source in the Universe. Some sources are closer and others more distant, but this radiation has been traveling through space since the emitting bodies were born. The result is that space is full of radiation, which, added to the time needed for interplanetary and interstellar travel, makes these missions today suicidal.

The radiation dose for nuclear power plant workers is 20 millisievent (mSv) per year, while for nuclear power plant accident liquidators it is 200 mSv. After a one-year stay on the ISS

- International Space Station - an astronaut receives about 220 mSv. That is, he receives more or less the dose of a liquidator.

Astronauts usually stay up to six months on the ISS, with occasional exceptions as we have seen (Scott Kelly). As exciting as it must be to be an astronaut, it seems right now that it has just lost a lot of its appeal. Radiation gets into our bodies, it is powerful, it breaks fundamental biological bonds and processes, which, in most cases, precipitates into cancers.

Possible problems for offspring should also be mentioned. Our ovules and sperm may be imperceptibly affected, but then passed on to our offspring with undesirable results.

Here on Earth, we have a weak shield, but enough to defend us from most of this radiation. The Earth's atmosphere and magnetic field are our weapons against cosmic radiation. That radiation wandering from space and on a trajectory towards the Earth almost always ends up trapped in the trap of our magnetic field, and diverted towards the magnetic poles, thus also providing us with the spectacle of the auroras, which is basically the radiation diverted towards the poles and colliding with the various atoms of our atmosphere. In an interplanetary or interstellar trip, we have neither magnetic field, nor atmosphere...

Illustration 6 - Earth's magnetic field

Apparently, a trip to Mars is still possible, but if to the effects of weightlessness and reduced gravity, we add these doses of radiation, the heroes returning from Mars will not be fit for parades, but rather for the ICU (intensive care unit) and let's see if we are lucky and get out of this...

This is under normal conditions. If we are lucky enough that our sun does not want to give us a solar storm. Solar storms are events in which solar activity is altered. It throws plasma and a much more intense solar wind into space. The solar wind is a stream of charged particles released from the solar corona. The plasma consists mainly of electrons, protons and alpha particles with thermal energies between 1.5 and 10 eV. As we have seen, the Earth's magnetic field is the natural shield that protects us from all this radiation, but in the processes of more solar activity, this magnetic field of ours is also altered, it is as if it is reduced. It is like when we raise our hands to protect our sight from the Sun. Now let's imagine that this sunlight is so powerful that it makes our arms bend. The same it is with the Earth's magnetic field. And if it is able to do that with the Earth's magnetic field, what could it do to unprotected astronauts on an interplanetary or interstellar journey?

But not only the Sun. Any violent event "nearby" in the Galaxy or in the Universe could have fatal consequences for the unprotected human body. These solar storms we are not in a position to foresee. We believe that there is some cycle around 25 years, but apart from that, other storms arise from time to time whose pattern we have not been able to discover at the moment.

In this situation, the reality is that astronauts on a trip to Mars have a real chance of being unlucky enough to suffer a solar storm with very harmful effects on their health. Let's keep in mind that there are no hospitals in space. Obviously, on a mission to Mars, one of the astronauts must be a doctor. But, even if this doctor has all the necessary equipment, in the face of major radiation, from which the doctor will not be spared either, one of the few treatments would be to apply morphine while waiting for the visit of death, which always appears as the last unwanted guest at the radiation party.

As I said at the beginning of this section, there are many problems with space exploration, but if we want to take this issue seriously, and we had better do so, we must begin to look at how to solve the problem of gravity and radiation.

The matter is serious since the existence of the Earth and the Sun has an expiration date. At the moment, distant date, but that does not mean that we should do nothing. It is like our life. Few thinks about death, the meaning of life, so that when the time comes, at least we have a clear idea, right or wrong, but with which we arrive at that moment calm. It is like when we are students and we have exams. You can study and arrive at the exam calmly, or study the night before and arrive at the exam nervous and crossing your fingers. Our life is also an exam for ourselves and it is never too late to start doing our homework. The same with the situation at hand, it is never too late to start thinking about how we are going to solve the problem of gravity and radiation, so that we will be able, when the time is right, to leave our home in the Solar System, before it becomes unstable and kills us.

So, let's start with gravity. How can we get gravity? By being close to a massive body, through centrifugal force or also with linear acceleration.

A rotating ship will produce the sensation of gravity inside its hull. Rotation displaces any object inside the ship toward its walls, giving the appearance of a gravitational pull directed outward. The thrust, known as centrifugal force is actually a manifestation of objects inside the ship attempting to travel in a straight line due to inertia. The walls of the spacecraft provide the centripetal force required for the objects to travel in a circle. The levels of artificial gravity thus achieved vary proportionally with the distance from the center of rotation. With a small radius of rotation, the amount of gravity felt over the head would be less than that felt at the feet. So, we would need a large radius, which is equivalent to a large spacecraft. It could also be avoided with slower rotations, which are also necessary to avoid as much as possible the "discomfort" that the Coriolis effect would produce. According to some studies, to reduce the Coriolis force to habitable levels, a spin radius of 2 rpm or less would be needed. To produce artificial Earth gravity with this rotation speed, we would have to consider spacecraft with a minimum radius of 224 meters. That is, a donut-shaped spacecraft with a radius of 224 meters, or a diameter of 448 meters spinning at 2 rpm. That's a huge structure spinning at a significant speed. If we were to spin the ISS - the International Space Station - at that speed, it would soon disintegrate. It would have to be a doughnut whose structural skeleton and walls could withstand those stresses. But let's assume that the engineering problems will be solved sooner or later.

Through constant linear acceleration, we would also achieve gravity. This idea in turn requires a constant supply of energy. While it is true that travel would be faster with this system, the energy input with the current state of development means putting huge amounts of fuel into orbit. As before, let us assume that it is a mere engineering problem that sooner or later will be solved, that there will be other types of engines with fuel, perhaps more complex, but much lighter for transport.

A gravity generator by magnetism is also being studied. It requires magnets with extremely powerful magnetic fields, involving expensive cryogenesis techniques and/or a lot of energy. In addition to the limitations regarding the use of ferromagnetic materials, it is not clear that humans could live in such an environment.

In any case, we see that the problem of gravity has theoretical and even practical solutions at small scales. What we also see is that there are some important engineering problems whose solutions and fruits we are still far from enjoying, but we must assume that, with time, these problems will be solved.

Once we are able to generate artificial gravity, we will also be able to modulate it and gradually adapt it to the gravity of the celestial body we are going to visit for both short and long trips, so that when we arrive at our destination, we will be perfectly accustomed to the local gravity.

Let us now see how we can solve the problem of radiation. At the moment, there are only two ways to preserve life. The first and simplest is to get into constructions, if possible, of water, hydrogen, plastic or metal, and the thicker the walls, the better. This solution does not seem very effective for a spacecraft, since most of its mass would be those heavy walls, and transporting and propelling them would not be an easy task.

The second method would be to copy nature, that is, to create a magnetic field around the spacecraft. Just as the Earth's magnetic field protects life on Earth, a sufficiently powerful magnetic field should also protect life on the spacecraft. Like artificial gravity, I understand that we would also be able to modulate it according to needs, or rather, emergencies. If we detect that we are approaching a barrage of particles from a solar storm or any other event, it would be enough to increase the power of our magnetic field to protect us.

So how do we create magnetic fields? With natural magnets, or with temporary magnets created by the circulation of electric currents. In other words, our ship should be an artificial magnet created by the circulation of electrical energy.

At the time, when I put these two problems together in my head, my head came up with the idea for such a spacecraft capable of producing artificial gravity and artificial magnetic fields at the same time. I would not say that it is my invention, but I proclaim that the genius of putting such an invention in space is mine. It is an alternator. I will not develop this idea here since the engineering problems are beyond me at the moment, but I will outline the main lines. This would be a basic diagram of an alternator.

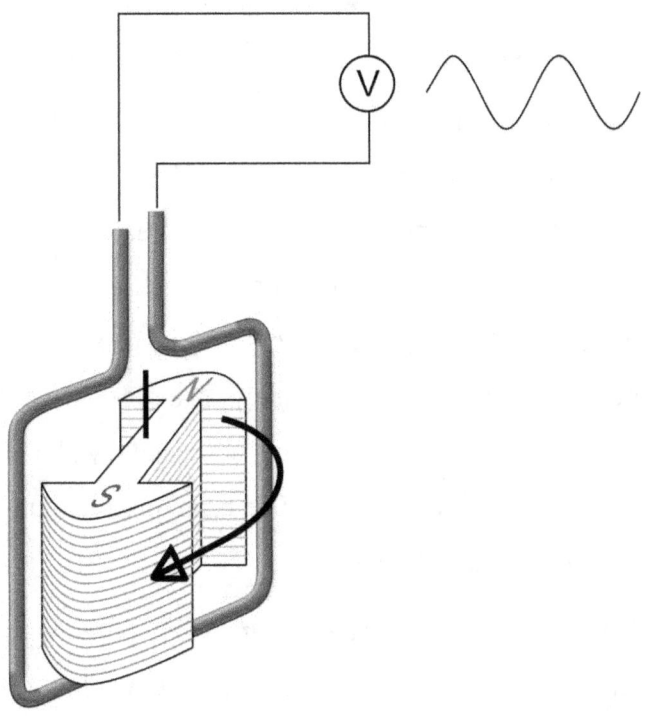

Illustration 7 - Diagram of a simple alternator with a rotating magnetic core (rotor) and stationary wire (stator) also showing the current induced in the stator by the rotating magnetic field of the rotor.

An alternator is an electrical machine capable of transforming mechanical energy into electrical energy by generating an alternating current through electromagnetic induction.

An alternator consists of two fundamental parts, the inductor, which creates the magnetic field, and the armature, which is the conductor crossed by the lines of force of the magnetic field.

Looking for a common resemblance of our lives, without having to know these things, we have the alternator (wrongly called dynamo), that some bicycles carry to create light through the movement generated by the wheels of the bicycle. Nowadays, there are even more modern light generators for bicycles that do not even touch the wheel, which is an advantage in order to avoid wear on the wheel of the bicycles. They are activated by the passage of magnets every time a wheel turn. The principle remains the same, although applied in a different way.

Both inductor and armature are solid on a common axis, and it is normal for the inductor to rotate while the armature remains static.

Let's see how we can apply this idea of an alternator to our spacecraft.

Recall that one alternative to our spacecraft to be able to produce artificial gravity, had been left with a shape of a torus, that is, the shape of a donut 448 meters in diameter, and that it was spinning at 2 rpm.

As mentioned above, in a typical alternator, the part that rotates is the inductor, i.e., the internal part, but I do not see why it cannot be the other way around, i.e., the inductor remains fixed, and the armature, the external part, rotates.

Now let's imagine that our 448 meters diameter donut rotating at 2 rpm is the armature in this alternator where what rotates is the armature, the external part. We need, therefore, a core, an inductor to complete the alternator, which should also remain static at the level of rotation with respect to the donut. It could also be used as a habitable habitat without gravity, storage or other major utilities that we will be able to get out of it, apart from its own and fundamental inductor in our spacecraft.

All this we should join it somehow so that both parts have the same axis. It can be done physically, through mechanisms, bearings and others, which are problematic and require maintenance, or, on the contrary, we could try to make this union virtual. This virtual joint would be constantly monitored and corrected through small thrusters and the ability of computers and sensors to keep both parts maintaining their constant relative positions with respect to the axis.

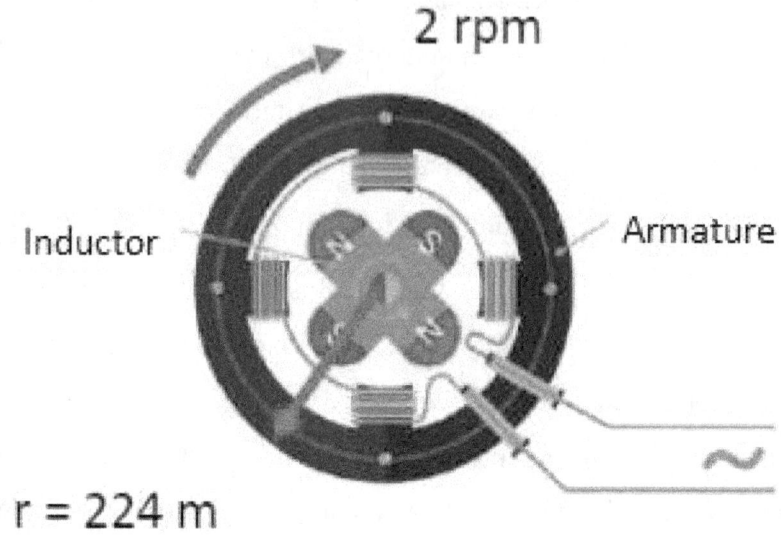

Illustration 8 - Spaceship prototype

In this system the electricity produced could even be considered as a secondary production product, of no interest, but which we would obviously use avidly for our own consumption of the habitat in the donut, or as a secondary source of power supply for electromagnets. Here, in order to avoid a continuous consumption of electricity for these electromagnets, we should rather think of permanent magnets, obviously artificially constructed and of the characteristics required by the detailed studies of this approach.

Around this entire ship, we would therefore have an artificial magnetic field. However, this should be sufficient to deflect most of the radiation, as well as to allow life inside the donut, so perhaps other fine adjustments would have to be made inside the donut so that the magnetic field would not be as strong inside the donut as in the surroundings and outside, in order to avoid possible harmful effects of this magnetic field on the health of humans.

This could be a first version of an interplanetary and/or interstellar spacecraft with enough features to avoid the two major problems of these trips, which, as we have said, are gravity and radiation.

A step further, but requiring technology completely out of our hands at the moment, would be the Death Star type ship from the movie Star Wars. This is a huge ship the size of a satellite or a small planet. The possibilities of creating artificial systems on ships of this size are multiplying. However, the principles should be the same.

Between that first model of ship and this second one, or even similar to the first model, we should think about the capture and adoption of small celestial bodies such as asteroids, to convert them into spaceships. The natural camouflage of these asteroid-ships would be ideal to attack by surprise the aliens who do not want to pay the tribute that by divine design corresponds to all of us earthlings... ☺.   It would be a matter of looking at the technological status, on the one hand, and making a study of asteroids that serve our purposes.

Some characteristics that these candidate asteroids should have would be the following:

- Some symmetry. The idea is to install motors of some kind to that asteroid so that it moves where we want it to go. In order to have easier and cheaper control, the more symmetrical the better. Otherwise, the lack of symmetry will cause that, for each impulse, we will have to correct trajectories, as well as to avoid that the asteroid enters in unwanted rotations, having to add new engines in locations to be specified;

- Natural magnetic field. If these candidates already have a magnetic field, it would be very positive. It would probably have to be enhanced and/or corrected, but it would be a very good seed;

- Water. If they have water in some form, we would have a very important store for our life, experiments and even protection;

- Strategic materials. Depending on the technological era, some materials are more useful than others. In any case, if there is a sufficient diversity of materials, they will be materials that will not have to be brought in from outside, and that can be used to manufacture what we need; and

- Gravity. In the bodies we are considering, the more gravity, the better, since it is a parameter that in an asteroid seems fixed and that we will not be able to vary.

Basically, it would be a matter of taking the right rock, put some propulsion engines, a little house, and travel. The little house I understand could be a combination of an outer dome plus a cave excavated inside the asteroid, where for protection is where most of the activities would take place.

Further, if we combine the idea of the Death Star and the asteroid ship, we could think of converting not only an asteroid, but an entire planet into a ship. Of course, the habitable zone of that planet-ship would have to be underground, since in case an atmosphere still persists, the continuous change of energy that it would receive from the various stars, perhaps cancelled natural rotation and others, would make that possible atmosphere, something completely chaotic and destructive. In any case, placing a propulsion system on a planet does not seem to me to be a simple task at this time, however, it would be the ideal ship.

In any case, as mentioned above, the human being as such will have to move out sooner or later and leave the Solar System before our star becomes unstable so that we are not imprisoned for extinction without a trace, except for a few fragile probes drifting through space.

22.   Point of view.

Let us now look at all this from another point of view. Let us assume that extraterrestrial civilizations exist, as it should be from a statistical and probability point of view.

This possible civilization locked in its planet, in case it decides to make interplanetary and interstellar trips, will face the same problems as we do, that is, gravity and radiation issues, as well as distances and times not common to any planetary scale.

It is true that these extraterrestrial beings may have a half-life of 1,000 years, but the distances, as we have seen, are still important.

So, let's go back to a question already raised: Have we already been visited by an extraterrestrial civilization? As always, there are probabilities, but I would say that at this present moment of the life of the Universe, of the Solar System, considering the distances, travel times, inherent problems of these trips, etc., etc., those probabilities should be very close to zero, considering the age we are giving to our Galaxy and the entire Universe (it is estimated that the Universe as such has an age of 13.8 billion years, while the Milky Way, perhaps 13.5 billion years).

As we have seen, everything is up in the air and depending on the value we give to certain parameters, we may be able to make extraterrestrial civilizations more common. Maybe life is more common than we think. Maybe clusters of civilizations emerge in somehow privileged galactic zones. All this is positive to give us hope that we are not alone, but they are only speculations to which we can give the degree of subjectivity and objectivity that we consider appropriate.

In view of all the above, let us imagine that some extraterrestrial civilization has succeeded. That is, it has managed to cross the distance of time and space between its home and ours, and the descendants who embarked on that ship are still alive, reproducing and evolving in that traveling home of theirs.

If they have been traveling to us for so long, and assuming they have more than enough technology to transmit communications, wouldn't they have contacted us by now? Showing up at someone's house uninvited and unannounced is not polite.

Since that signal or signals, and considering that there could be several civilizations traveling towards us, have not yet been produced, we must give some explanation for this lack, some of which I proceed to expose:

1. There is no communication, because such communication has never taken place. That is, there is no civilization except our own.

2. Such communication has not yet reached us. As we saw in previous chapters, even communication at the speed of light takes time to travel from one star to another. So, someday we will receive a signal from some civilization, which may also be on its way to our Solar System. It would also be possible that we establish such communication, but that this possible civilization is not interested in visiting us, because we are not interesting enough, or because our star does not have much life left, etc.

3. Such communication has occurred in our past when we were not yet ready, and by the time we were ready, that civilization has already been extinguished. We were either too late or they were too early.

4. Such communication has already occurred, but within some conspiracy theory, the governments and scientists involved are keeping the secret for hidden purposes (and all of them related to all our openings) for their own benefit so that we do not panic.

5. Extraterrestrial civilization is not interested in communicating with us because:

    a. We are considered a backward and uninteresting civilization. It's as if we went to a zoo and they showed us elephants, rhinos, tigers, orcas, sharks, whales, and then some worker ants making tunnels. Interesting, but

I'd rather see a shark. We would be those little ants without much interest.

b. They want to conquer us. They are going to steal our planet (which, let's remember, doesn't belong to us either...), to use all its goods, and to use us as cattle in the best-case scenario. But most likely they will kill us all.

c. They want to sexually abuse us. After a tedious journey of thousands of years, they have finally arrived on Earth, and they imprison earthlings in sparsely inhabited wastelands to satisfy their repressed sexual desires during those long years of travel, with the attractive earthlings. They insert their sexual organs, their fingers, probes, dildos, surgical and musical instruments through each and every orifice of these attractive terrestrials in a galactic orgy. And if they don't have enough holes, they open some more, and then they carefully plug them up, leaving unmistakable marks on the bodies of these abducted and sexually assaulted earthlings.

Let us bear in mind for this last section that it is practically impossible to block all types of communication. Even if no one is deliberately writing to us, if there is a civilization capable of doing what we are considering, we should be able to detect their communications, even if they are not directed to us. Let us remember that we ourselves are continuously transmitting thousands of communications, television programs, series, movies, news, etc., into space. By this I mean that it is very unlikely that they will catch us by surprise to attack us, let alone rape us, so point 5 of these alternatives has very little real chance.

Now let's note that all UFO sightings supported would fall under point 4, conspiracy, or subsection 5c, sexual assault, or any of its variants. Without going into detail on those variants of point 5c, I proceed to comment on sexual assault to show the ridiculousness of this type of thinking. How is it possible that extraterrestrial beings after their feat of interstellar travel hide their presence only to rape earthlings!?...

It is possible that our Solar System does not have an optimal location within the Galaxy for this type of communications and even visits to have occurred. Our beloved star is in the outskirts of one of the galactic arms, as if we were in a peripheral and even marginal neighborhood of a big city. If there are other civilizations and they are in contact, galactic routes will have been created, promoted by trade, knowledge, wars or any other possibility, or a combination of all of them. In this framework, our location in the

Galaxy is not good. But if these routes existed, then even though we were not part of them, we should be able to detect them.

This again leads us to the conclusion that either they do not exist, or such communications have not yet reached us, or reached us before we would be ready.

If communications have not yet reached us, this means that at least in our Galaxy, the Milky Way, these possible civilizations are young civilizations, like ours, at least in the technological field.

Let's delimit a little more this concept of youth. As we said before, the Milky Way has an average diameter of 200,000 light years. Since we are in the outskirts, we can take this data as the maximum, thinking that a possible civilization is diametrically opposite to us (it would also be bad luck...).

This means that if a technologically advanced civilization in the other part of our Galaxy started emitting 200,000 years ago, when the first modern humans were appearing on planet Earth, those emissions would now be just arriving. And that would be the maximum limit. That is to say, between those 200,000 years and today, any possible communication within our Galaxy would be covered.

200,000 years for us who normally live less than 100 years, is an outrage, but on evolutionary scales of the Universe, it is not much. As I said, 200,000 years ago, modern humans like us already existed in some places on Earth.

That 200,000 years within the age of the Milky Way at 13.5 billion years is 0.0015% of its history, which is a ridiculously low figure.

Once again, we are obliged to give some explanations for this pathetic little number.

To begin with, any kind of life, which has also evolved technologically, needs heavy elements. In this context, any element of the periodic table with the exception of hydrogen, helium and lithium can be considered heavy elements. This requires that several stars of our newly formed young Milky Way lived and died to create heavy elements capable of creating in later phases new stars with planets and perhaps life.

Although there is a planetary system only 375 light-years away with an age of 12.8 billion years, it consists of planets with no heavy elements. So, when did the first planets with heavy elements form? Well, it is complicated to answer this question, but let's consider our Solar System with its four rocky planets as a first example and a first approximation. The estimated age of the Solar System is 4,568 million years, that is, 1/3 of the age of the Milky Way, time that has been needed for other stars to live and die producing heavy elements, as well as for the appropriate conditions for the formation of the Solar System to occur.

After birth, we have to consider some time for planetary stability and life to emerge. This is thought to have occurred on Earth 3,460 million years ago, for some elementary form of life. We found some complex life 2,000 million years ago, and from there, evolution through several mass extinctions to reach until us.

Well, seen from this perspective, ours and the only one we have at the moment, it is not surprising now the situation of uncommunication in which we are. 200,000 years versus 2 billion years for the first basic forms of complex life is 0.01%. Still a tiny number, but larger than the previous one.

The first intelligent being, and here we should first define what intelligence is, although we will leave it for the moment, arose when certain cells specialize and join together to form a kind of first brain ganglion in some mollusks. Something like octopuses. That was 400 million years ago.

That first intelligence has to evolve and develop sufficient technology. Let us note that octopuses as such, or any mollusk in general, although it seems that they were the first to have some intelligence, are not the life form that dominates the planet, and worse, although they are intelligent, their evolution from those 400 million years until today in comparison with other species and everything that has happened around them, leaves much to be desired.

Then we must add, it seems, other ingredients to make the necessary leaps. Those ingredients have to be natural, so we only have to look at the environmental pressures of the planet, as well as the episodes, some fortuitous and some not, of mass extinctions, which forced living beings to evolve as they did, and not in an alternative or parallel way.

As we can see, it is a whole chain of evolutionary facts dependent on each other that are able to take us from the very beginning to the present day, and give us a more or less coherent explanation of why we do not have communication with other civilizations.

The conclusion could be that the chain of precise conditions for the emergence of a technologically advanced civilization takes time, time that should be comparable to what the human being himself has needed so far.

This in turn means that other civilizations may have existed before ours, but if we no longer see them, we must assume that they have become extinct, or that they never existed.

In this context, we must also assume that all possible civilizations are "young" in the terms of time we have been discussing.

Finally, we should also mention that it is quite possible that we are among the first technologically advanced civilizations to have emerged in the Milky Way, and it is also possible that at the present date we are the only one.

Since the age of the Universe and that of the Milky Way is similar, these conclusions should be applicable to all galaxies. The only difference is that if there are other civilizations in other galaxies, as we have seen in previous sections, the times are much longer, and obviously, with the distances, communications are diluted and weakened making them perhaps undetectable because they cannot be separated from the background noise.

Evidently, the accumulation of circumstances that have made our existence and our technology possible, may occur in other locations in a multitude of ways and perhaps in a much more accelerated way, but although it is a possibility, given that we have no record of such a fact, we should discard it for the moment.

It is also possible that the chain of factors that cause civilizations with technology may take much longer, or even be very isolated cases, among which we would find ourselves.

Again, and in spite of being on a planet around a common star in a typical spiral galaxy of a rather vulgar local group of galaxies somewhere in a monumental Cosmos, we must consider ourselves special, and with this to revalue in its just measure the life of any living being on this planet, as well as our beautiful planet.

23.   Specials.

Also, the special circumstance in which we find ourselves makes me think again about the simulation theory and whether all this is nothing more than a simulation or perhaps a dream. The special circumstance I am referring to would be the following. We are a civilization with a sufficient degree of biological and technological evolution to raise all these questions. For example, about the existence of other civilizations. Our conclusion is that there are probabilities, but everything seems to indicate that we are just in the epoch of the life of the Universe where these civilizations could exist, but whose existence would be recent, perhaps so recent that any type of communication has not yet been possible. That is to say, it seems that we are always there, on the verge of something, it is an almost, almost, that never seems to come to fruition.

It all seems like a door ajar inviting us to pass, but when we try, we are too fat to pass, and the door won't move and we don't know how to move it. Our dream can sometimes be a nightmare.

So, have aliens visited us? I doubt it very much, let alone to abuse us. In a possible visit, they would be detected a long time in advance, we would start communicating with them before their arrival for tens, maybe hundreds of years, and when they arrive, it would be a reception at planet level, not country or some countries, in a big city agreed by all of us and I understand that it would be a meeting where the forms would be impeccable on both sides since we would already know each other after the long years of waiting.

Small flying UFOs are not capable of intergalactic travel, larger ships are needed, motherships with all their detectable paraphernalia.

Faced with that we can talk about travel faster than the speed of light, but since this is beyond our knowledge and, in fact, based on our knowledge is something that is not possible, I will not consider it. It would be great, yes, but it is not possible (at least for now and with current knowledge).

It could also be that we have ships where all travelers are frozen during the journey, and the ship is directed by computers. These computers would perform the necessary maneuvers of deceleration and approach, without emitting any communication, and perhaps they would pass unnoticed to our telescopes. Then is when the sleepers would wake up and if there are enough of them, they would wipe us out, and if there are few of them, they would only rape a few of us. No doubt this is possible, but it would be a rather pitiful life form in which case perhaps better to be annihilated (I do not consider the other possibility).

I mean that these interstellar journeys are journeys of no return. On a large scale they should be journeys to save civilization since the host star always ends up dying, and to our planet, it will also happen with the death of the Sun. This will happen, and we must assemble a ship capable of taking us in and start the great journey of no return.

Now, we may not all be able to go... And those who do go, will they voluntarily want to be frozen and perhaps never wake up? And if they do wake up, will the first thing they want to do is war and rape? Well, on this occasion I must contradict myself to some extent, since we may have to answer this question at least partially in the affirmative.

From what has been said, these interstellar voyages, depending on the distances and the remaining life of the star that has allowed the birth of a technological civilization, could be more like Noah's arks than commercial and/or diplomatic missions.

If this Noah's Ark is heading towards a pre-fixed destination where some kind of communication already exists, I understand that it will be a friendly encounter between civilizations, but, even so, let us remember what our history tells us. Let us remember what happened when the Europeans met the Americans. Both the Europeans and the Americans were human civilizations where by fortuitous circumstances, geographical, directional, etc.... the

Europeans were able to develop a technology that in 1500 was superior to that of the Americans. The circumstances could have been different and it would have been the Americans who entered Europe. In any case, the clash of civilizations occurred, and despite the efforts of many to have a happy encounter where wealth was shared, the net result was the practical annihilation of the American people, and subsequent enslavement of part of Africa. And yes, perhaps the Spaniards started it, but all the European states that were able to take advantage of this encounter, did so, and each one bloodier than the previous one. Because here I am obliged to say that, in the South of America, the Europeans left at least some survivors, but in the North, the Americans were systematically annihilated until recently by the Europeans who came from a great European island, and even today the pitiful remains of those ancient Americans of the North, are kept marginalized in "reserves", lands that nobody wants...

In addition to their own religiously motivated confrontation with a great promoter, not divine, but economic to steal all the American riches, we must not forget the even greater havoc wreaked by viruses brought from Europe for which the Americans had no antibodies.

Having said that, the peaceful encounter between a possible interplanetary Noah's Ark and a planet inhabited by another civilization does not look too good at first. It is very likely that, after a first happy and peaceful encounter, everything will twist gradually or suddenly, where someone will finally have to come out as the winner, although in the course of becoming the winner, will have to lose a lot, including its essence, and the other party will be the vanquished if there is anything left of that part, otherwise it will become the annihilated party.

But now let us suppose that this interplanetary Noah's Ark begins its journey out of necessity in the face of the imminent death of its star, and without yet having a predetermined destination, since it has not yet had any communication. With the perspective enunciated above, and in the face of the unknown, would this not be a militarized Noah's Ark, ready for any possible confrontation? Perhaps it is the human mind of mine that makes me think this way so that our genetics will continue to be passed on to the next generation, but I am afraid I have no choice. Or do I? What was it about contemplating the Universe as a whole? That thing perceived in dreams and meditations?

Two do not fight if one does not want to, and the one who does not fight has to be patient for the opponent to understand the whole picture. It takes time to understand each other, to see benefits and detriments. Once both parties are in the same sphere of understanding, it is unlikely that there will be confrontation, but it takes time, patience, and obviously the non-fighting party, while waiting and trying to explain himself, also has to defend himself if necessary.

We could say that avoiding confrontation is cowardice, but you have to be very brave to swallow your pride and run away. If there is no other way out, one fights, and fights brutally, without regard and until a total victory by totally annulling the enemy. But if there is a way out, it must be taken. That would be the intelligent attitude where our mind must overcome our temperament with boiling blood. It is better to run away if a confrontation is foreseen, than to stay and fight, where, even if you win, you will have lost. Physically and psychically, you are no longer the same person. No one is the same after a war, and if not, ask the veterans. Besides, the proposal to run away that I propose would be more like one of keeping the enemy at a distance, but trying to talk to him. It would be something like one of the scenes from the Monthy Pyton's "Life of Brian" movie. The scene I am referring to is when a poor wretch in basically a loincloth is thrown into the arena of a Roman circus where a gladiator armed to the teeth awaits him. Here the wretch decides not to engage and starts running. The gladiator behind him, but with the added weight of his armor and weapons, he immediately falls to his knees, while the wretch looks like he could still run a marathon. Thus, the proposal in the face of a confrontation between civilizations would be to keep our distance until there is a deep mutual understanding of the common position of both sides in the Universe and the meaning of our own existence.

Where 2 eat, 4 can eat. Although people die of hunger on our planet, this should not be the case with good governmental management at the planetary level, and is perfectly avoidable. But countries insist on unbalancing the balance of equality to have more during their ephemeral lives by stealing from others, drawing borders on maps where there is no natural border. Our ape brain has yet to evolve enough for us to see the futility of these historical wars of ours. I hope that in another 200,000 years we will see this time in our history as a turning point when Homo Sapiens finally became true sapiens and saw beyond the banana in front of their eyes embedded in a basically ape brain.

It is also possible that, in our possible journey in search of a new home, we find a planet that is habitable for us and where there is no civilization of any kind. In that case, there would be no confrontation, at least for us, and we would take the planet completely keeping our super interstellar ship ready for the next trip that would occur again with the proximity of the death of the host star.

That possible habitable planet, but without civilization, may have some forms of life, which would be annihilated as a virus and before any possible threat to us, or enslaved for our use and enjoyment. Basically, if we find a habitable planet, but without a civilization, we will take the whole planet and steal its future and all its possibilities, to make it part of our history.

The future of the human race is to assume the role of interstellar wanderer in search of temporary homes, perhaps in the company of other beings, perhaps alone.

Let us return for a moment to those militarized Noah's Arks. It is more than likely that in that Ark we will not all of us. There will be the rulers, the monetarily privileged or those with power who in turn have placed the rulers in their armchairs, and those people who are considered necessary on their own merits in various fields such as medicine, engineering, natural sciences, etc.

The rest of us, I fear, will be left without a planet and without a future.

That representation of mankind tucked into its ark and searching for planets will suddenly have managed to bypass its programmed end despite the price of leaving millions of fellow humans waiting for their end. Equally, leaving someone behind in this approach means brutally cutting back the gene pool of humanity, depriving the surviving travelers of incalculable evolutionary possibilities. This obviously has the clear danger that the surviving humanity will end up degenerating, or at least not progressing. To all this, we must add the rigors of the travel.

Of course, by then genetic manipulation will be an everyday occurrence and even the moral barrier within an ark survival framework may disappear completely to the point of not remembering that it ever existed.

Here again we must somehow integrate Artificial Intelligence and see the relationship with humanity, and their evolution together and/or separately.

Whether we get smarter and smarter or perhaps dumber and dumber, whether we are humans with more and more complements, whether we are half-human - half-machines, or whether we are only machines, the legacy of organic beings that have already disappeared, the future remains the same: wandering from one planet to another, which will be increasingly difficult, until we finally dissolve with the Universe in its accelerated expansion.

24.   Conclusions?

Trying to answer the question of whether we have already received the visit of some extraterrestrials, we have arrived here, where we foresee that, if we manage to survive beyond the dictates of the Sun's own life, then we will be the aliens in search of a home, with a future that, although distant, does not cease to be equally certain and depressing, the total disappearance. Just like our own death, but on a global and generalized level, the death of humanity. Which in turn reconfirms that we must be brave and face life fully. To give meaning to our life is to give meaning to humanity, each one through his mission: to love, to guide, to have offspring, to write a book, to plant a tree, ... There is no big or small mission, the important thing is to complete it or at least try, but really try. The rest is hypocrisy and waste of life. We are all the same, rich, poor, poets, beggars, we all shit the same, and we all die the same, alone, even if we are surrounded by people and relatives, there are roads that one must do alone, where there is no one to guide you, protect you, or behind which you can hide. Do not let yourself be blinded by the lights of this modern world of ours, be brave and take your path alone, because once you walk alone, you will see that you are really accompanied by all those who have also dared and dared to walk alone. Thus, when the ultimate end comes, both on an individual level, as well as on a human level, we will not be afraid, since it is another path, but similar to others that we have already walked. Walk alone to be in good and true company. As some walker said, it is normal that we die, but we only do it once.

Everything can be taken away from us, everything, even life, there is only one exception, the freedom to choose, the attitude with which we face any circumstance regardless of the conditioning factors. It is about playing the game, and getting the most out of that game, or at least trying to do so, whatever the cards you have to play with. We do not control what happens, we control how we respond. We must mature and stop this childish behavior that haunts us in which if we do not get good cards or if we see that we are not going to win, we get angry and do not play. Not playing your game is to despise your life and your opportunities, no matter how few they may be.

You can pretend to have other cards, but then you will be playing someone else's game, you will only be an actor playing a role and even if you reach your deathbed with that mask on and no one has discovered you, you will continue to live in a lie because you know the truth. You know that all your life you have been a coward pretending to be someone else because you didn't like the cards you were dealt.

When you play your cards, you are not afraid of life or death. One perceives as in meditation and sometimes in dreams that there is something more, a connection of the whole with the whole.

25.   The "I" is dead.

And I take advantage of these last lines to change the subject completely, again. Let's talk now about what we call conscious being, about that which we call "I". Here I have to make an announcement, which, although it is not new, remains hidden, because I understand that it is something difficult to digest and it is "better" to ignore it as if no one had already said it before. The news is that what we call "I" does not exist. Buddhists said it a long time ago, and they discovered it through their meditations. Nowadays, medical studies confirm it. But, even so, it seems that we are still attached to that "I" inside our heads that do not stop talking to us. Let's begin.

In Europe, for centuries we have been dissecting bodies to find that "I", that consciousness or that soul. No matter how much we cut, we have not found anything like that. There are those who say that it is in the heart, when even though it is vital, it is still a muscular organ that, by orders it receives through electrical impulses from the brain, moves to pump blood throughout the body.

Others say that we should look in the encephalon, what we vulgarly call brain, in our head, and indeed it seems the right place, since it is the place where all the external perceptions that reach us through the five senses end up, where these perceptions are analyzed, and where finally and after this analysis, orders are given to the rest of the body to act. It is where our thoughts arise, where we recreate our memories and fantasies, where there is activity when we dream. Undoubtedly, that is where our "I" is...

However, if we dissect a brain in slices until we try to see everything, we only see basic neurons as bricks of that building, but there is not a sort of black box where our "I" is enclosed. Moreover, in those living beings where, due to various circumstances, their corpus callosum has been severed, it is not that there is one "I", but that there are 2 "I's".

The corpus callosum is the largest bundle of nerve fibers in the human brain. Its function is to serve as a communication pathway between the right and left cerebral hemispheres, so that both sides of the brain work together and complement each other.

People born with agenesis of the corpus callosum, that is without it, present neurological problems, syneresis and comprehension problems, since their mind works as if they had a split brain. That is to say, it would be functionally like a person with two brains, since the information received only by one of the hemispheres would not pass to the other.

There are also people who were normal at birth who were intervened to cut the corpus callosum, due to an accident, or for example to reduce refractory epilepsy.

There is extensive literature on the study and behavior of these people with their two brains, but what interests us here is that, in the absence of communication between the two hemispheres, the brain no longer works together. If it does not work together, it is divided and we can say that there are two brains. Two very similar brains since they belong to the same person, but differentiated. They don't even seem to care what their other brain thinks. Experiments have shown that, by limiting the input of information to one or the other hemisphere, the behavior is different. In this sense, where is the "I"? Where would the consciousness be? If we see that the right hemisphere is more conflictive and/or problematic, would it be legitimate to deprive it of oxygen so that it stops functioning and we are left with only the left hemisphere? Or would this be murder? But whose murder if there is still a person with a functional left hemisphere?

The unique consciousness, that "I" does not exist, it does not cease to be electrical processes within the biology of the brains.

Nor is it necessary to study these extreme cases. The "immutable self" evolves, it is not the same as it was in our childhood or 10 years ago. And as we have seen, if the corpus callosum were sectioned, this "I" would be divided into 2 "I's", since each of the hemispheres develops its own personality, which may differ more or less from the one next to it, but with which it does not speak, and as has been empirically demonstrated, two separate consciousnesses develop.

So far, so well known, and this is where my contribution comes in since I have not read anything about it.

The explanation for why our brain is divided into two hemispheres is that our body also has two symmetrical but independently functioning parts. I understand that this is where the answer may come from.

Now, if we have two hemispheres because we have two symmetrical parts, what would happen if we would have another configuration with more symmetries? Four arms, four legs, etc. ... Having four legs and four arms does not seem at a practical level too clear and perhaps evolution discarded this option some time ago. But here we are at a theoretical level: would we be talking about four cerebral hemispheres as well, with their corpus callosum? And if we sectioned corpus callosum, would we have four personalities inside one head? Based on what we know about our real encephalon of two divisible hemispheres both at the physical and conscious level, I understand that the answer would be in the affirmative.

And indeed, it does. An octopus has 8 legs, it has quadruple symmetry. According to the results of various research studies, octopuses remember, manipulate objects, play, and so on. The neurons of the nervous system of octopuses are spread throughout their eight tentacles, which process information autonomously, as if they had eight brains.

But this is not the point I wanted to get to. Let us imagine for a moment that what I am going to say now is possible, seen from any point of view, especially at the surgical level perhaps. As a starting point let us take the fact that our brain divided into two hemispheres can be physically divided by cutting the corpus callosum, which causes the division also of our consciousness, our "I".

What would happen if instead of cutting, we glued? ($\odot$__ $\odot$)

Let's look at this question in a first phase, which is not the one that interests me, but which may be clarifying posteriori.

That first phase would be to look at a person with a severed corpus callosum and his two brains. If we were able to unite their corpus callosum, communication between the two hemispheres would be restored and perhaps after a brief shock, the two consciousnesses would engage in a dialogue that would lead them to unite into one.

And this is the option that interests me, and it is the one I take, although I am aware that this dialogue between the two hemispheres could also give other results, such as, for example, that one imposes itself on the other and enslaves it in some way, or that both, due to the pressure and stress of the situation, collapse and stop working.

I take the license to assume that everything is going well in order to continue with my presentation.

Let us even imagine that the repair of the corpus callosum has not only gone well and we have managed to go from two consciousnesses to one, restoring the communication between hemispheres, but that, during the repair of the corpus callosum, we have also installed a kind of switch. A switch that allows communication between hemispheres when it is closed, and as if the corpus callosum would be cut with the switch open. By flipping the switch, we would have a human being with one or two consciousnesses, according to our convenience (although at the moment I do not see any advantage to this...).

Even with the switch closed, and therefore when there is communication, a monitored filter could also be installed to allow the communication of information that we were interested in and cancel the information that we would not want to exist between the two hemispheres, even here for medical purposes to avoid some kind of disorder, and not just as an obsessive manipulator as I am doing here.

This would be the hypothetical framework of the first phase in which we unite, separate and even filter the connection between the cerebral hemispheres in order to obtain one consciousness, two consciousnesses or a lame (filtered) consciousness.

Let's move on to the second phase, which is the one that interests me.

Let us imagine the above hypothetical framework, but transferred to people instead of brain hemispheres. That is, let's imagine that we can put two whole brains together, yours and mine, for example, or yours and your neighbor's, or any two other unrelated people's brains.

Let's imagine that we are able to enter any human brain through a kind of female plug installed in the back of our neck, like the one worn by the protagonists of the Matrix movie, or through another not so invasive method, for example, through sensors temporarily attached to the skin of our shaved head. Let us also imagine that we build a bundle of optical fibers that would play the role of a corpus callosum between cerebral hemispheres, but here, between two brains of two people.

Let's imagine that we already have it, two brains of two people in total communication. What would happen?

Open communication through such a channel would enable an uncensored and unprecedented exchange of information between the two people. The two people would share their innermost secrets, thoughts and emotions.

But my question is the following. If consciousness is mutable as it is demonstrated when the corpus callosum of a person is sectioned, or simply by seeing how the "I" of any person evolves over the years, how far can we go with the mutability of our consciousness, our "I"?

Let's see, it is clear that consciousness evolves with time. It is also clear that we can divide it, at least in two by sectioning the corpus callosum between the two cerebral hemispheres. If it can be divided, can it be added, or even multiplied? In a first phase I proposed the union of two cerebral hemispheres with a defective corpus callosum. In this second phase I propose to join two brains of two independent persons. After the open communication between two brains of two persons, would a new consciousness be created from the two previous ones? My answer to the above question would be in the positive sense, provided that this union is given the necessary time. Perhaps we could create a system of removal and replacement, and carry out this union in phases so that it would not be so traumatic.

Evidently it seems a dangerous game, since the separation I understand could be very painful with probably unwanted effects. If two brains come together, even creating a new single consciousness, and suddenly separate, it would be like being robbed of part of them, as if they were amputated brains in search of their other brain. You would have to get used to being alone again without such fluid communication with another brain. I understand that the psychological problems would be many and varied.

Therefore, and given that this undertaking seems arduous and difficult, and that, even if it were possible, I propose a filtered, not total, communication. It would be a matter of the fiber optic cable that connects brains passing through a computer that would filter the information that can and cannot be transmitted. It would be an exceptional way to learn and acquire knowledge. I join my brain to that of a Frenchman with a total filter except for the language. After a while, I know French like a native. I put my brain to that of a commercial airline pilot, with the relevant filters, and after another moment, I know how to fly airplanes. All kinds of knowledge could be stored in computers, and it would be enough to select learning modules and insert them into your brain. With this we return again to scenes from the Matrix movie...

But let us return again to the possibility of merging brains. While achieving a single consciousness from two is perhaps not something we should proceed with immediately because of the possible consequences, the theoretical possibility is there.

Let's go now to a third, fourth and a bunch of other phases all at once.

If we join two consciousnesses to form one, what prevents us from joining three consciousnesses to form only one? Nothing. What if we join more? What if we join the brain of a human with that of a cow? What if we join the brains of all living beings that have brains or something similar to what we understand by brain? Would we also have here what we could call the Super Consciousness of intelligent beings on Planet Earth?

It would be the utopian world sought. Since we would all be integral parts of that Super Consciousness, we would mourn the loss of any one part. This would lead directly to the end of all kinds of wars, to the sharing of wealth equally throughout the world just as a gas spreads evenly in a closed enclosure. We would all be vegetarians and there would be no animal exploitation as there is today, which in turn causes large amounts of pollution. We would have to see how we could stop lions from eating wildebeests or zebras, and convince them to eat lettuce, but that would be another story...

For the above mentioned with respect to interstellar travel and possible contact with other civilizations, this Super Consciousness would be limited for the moment to planet Earth, but if one day the appropriate circumstances arise for contact with other extraterrestrial civilizations, the path should try to be the same, to unite our consciousnesses to enrich ourselves with knowledge and to live in peace.

Now let's go one step further. All this that I am telling you will happen in a future to be determined and that right now seems like the plot for a science fiction movie. But if we open the focal point to have a wider field of vision, we will see that we can achieve this already today in a rudimentary and poor way, without having such a massive transmission of information as the one exposed, but glimpsing the end. How? Meditation and dreams. When we talked about meditation, didn't we say that a cosmic connection of the whole with the whole was perceived? Do not the masters of meditation insist on confirming to us that the "I" as such does not exist? Does not meditation make us see the rest of humans as brothers and sisters? Does it not awaken in us a strong feeling of compassion towards all living beings?

Directed and uncontrolled dreams can also reach these same conclusions.

So, don't these three paths lead us to the same result? Don't dreams, meditation and the possible union of brains tell us which is the path we should take? I believe they do.

So, what should we do now? Evolve. I am sure that in the future it will be possible to surgically join brains, but it is certainly not the cleanest way and the psychological consequences of proceeding in such a crude way seem many and problematic. However, meditation and dreams are within reach today and with today's offerings, they are free! The only thing that is necessary, is time, of the only real thing that in this life we have a limited and always diminishing capacity.

To begin with, we should meditate every day. Include it in our daily routine like brushing our teeth. Write it in our diaries to block out that time, "Meditation Time". Make it a habit. As the days go by, in theory reaching deep phases will require less time, but we may want to stay in them longer. Getting started in meditation and going deeper is an individual path that only you can do. It is one of those things that must be done alone. There are aids and courses, but no one will tell you what your inner world is like.

Another initiative would be to try to perceive the reality of our environment in our daily lives, without being influenced by our senses and/or emotions. Obviously, this is not easy, especially in stressful situations, but it is precisely in those situations when we need to see things as they really are, and not as we think they are, in order to make the right decisions. There are people who naturally have this ability, it is said that they are cold-blooded in situations where the rest of the world gets hot. We know that when our blood gets hot and starts to boil, it is as if our intellectual capacity goes through a funnel and comes out at the small end, turning us into little more than any other animal with an ape-like physiognomy looking for a way out. These are the situations we have to master, and automatically. Not only cold-blooded, but icy. To be stones with glued eyes, or better yet like a robot with its sensors. Someone will say, but that takes away what we are, our humanity! To which I respond in the negative. When your enemy appears before you, the only thing you will see will be another human being with whom for some reason you do not get along, if you have reached this point and have tried any of the practices mentioned above, you will see that you do not really hate your enemy, at most you will feel sorry for them because if your enemy still hates you, it is because they are in a lower sphere of knowledge than you. You will even want to help them to pass to your sphere and stop suffering and tormenting themselves with this enmity, you will have compassion for your enemy.

On the other hand, when a loved one appears before you, you will again perceive the same thing, a human being, with whom for various circumstances you get along well and who makes you awaken certain pleasant feelings. If you are in the same sphere of knowledge, you will have a permanent smile of complicity on your faces. And if not, again there will be compassion on our part forgiving his possible mistakes and weaknesses, trying to help again.

Meditation will make that in our daily life we really start to see and not just look. Persistent meditation should little by little and with the passing of generations, make our brain evolve in the sense that we want to finally be able to share consciousness.

On the other hand, dreams should not only be remembered, but we should also try to control them. I am a lousy novice in this respect, but something tells me that as with meditation, in this field we should also be able to achieve great goals. Being conscious in your sleep is the same as being the master of the Universe, everything is at your feet. Now let's imagine that we "meet" to dream together. I ask you: "Shall we meet tonight to see each other in our dreams? What million doors does this possibility open? Being together in a dream and being aware in turn that you are dreaming would be a very intimate experience. In the dream mode where anything goes, could we swap? Could it be you, and you, me, if only for an instant? Could we mingle, besides physically, which I take for granted, on a conscious level?

If we can't "meet" to dream together, I can recreate the whole world where maybe, if it's interesting, you are there too. But maybe I find it more interesting to dream a dream where I am Doctor Consciousness and my world is only populated by copies of myself. While my clones live, breathe, work, fight and so on, Doctor Consciousness, who is just another copy with a white coat and a kind of light on his head, keeps taking notes until he finally exclaims: "Eureka!", since he has just discovered that "I" does not exist and my copies begin to explode one by one like soap bubbles making a "puff" when they disappear. The last copy to vanish is Doctor Consciousness and all is deserted and silent. Then a murmur is perceived in the distance, but it is a murmur that approaches and gets louder and louder. It is an overwhelming murmur, it is already a deafening murmur, and that which provokes it is already here. It's you, it's me, it's my mother, it's thousands of birds and insects, it's all of us, it's me.

www.ingramcontent.com/pod-product-compliance
Lightning Source LLC
Chambersburg PA
CBHW070328220526
45467CB00001B/77